101 KEY IDEAS

CHEMISTRY

Andrew Scott

D1315525

TEACH YOURSELF BOOKS

For UK orders: please contact Bookpoint Ltd, 130 Milton Park, Abingdon, Oxon OX14 4SB. Telephone: (44) 01235 827720. Fax: (44) 01235 400454. Lines are open from 9.00–6.00, Monday to Saturday, with a 24-hour message answering service. Email address: orders@bookpoint.co.uk

For U.S.A. order enquiries: please contact McGraw-Hill Customer Services, P.O. Box 545, Blacklick, OH 43004-0545, U.S.A. Telephone: 1-800-722-4728. Fax: 1-614-755-5645.

For Canada order enquiries: please contact McGraw-Hill Ryerson Ld., 300 Water St, Whitby, Ontario L1N 9B6, Canada. Telephone: 905 430 5000. Fax: 905 430 5020.

Long renowned as the authoritative source for self-guided learning – with more than 30 million copies sold worldwide – the *Teach Yourself* series includes over 300 titles in the fields of languages, crafts, hobbies, business and education.

British Library Cataloguing in Publication Data
A catalogue record for this title is available from The British Library.

Library of Congress Catalog Card Number: On file

First published in UK 2001 by Hodder Headline Plc., 338 Euston Road, London, NW1 3BH.

First published in US 2001 by Contemporary Books, A Division of The McGraw-Hill Companies, 1 Prudential Plaza, 130 East Randolph Street, Chicago, Illinois 60601 U.S.A.

The 'Teach Yourself' name and logo are registered trade marks of Hodder & Stoughton Ltd.

Copyright © 2001 Andrew Scott

Cover illustration by Mike Stones
Typeset by Transet Limited, Coventry, England.
Printed in Great Britain for Hodder & Stoughton Educational, a division of Hodder Headline Plc, 338 Euston Road, London NW1 3BH by Cox & Wyman Ltd, Reading, Berkshire.

Impression number   10 9 8 7 6 5 4 3 2
Year   2007 2006 2005 2004 2003 2002

Contents

Introduction

Welcome to the **Teach Yourself 101 Key Ideas** series. We hope that you will find both this book and others in the series to be useful, interesting and informative. The purpose of the series is to provide an introduction to a wide range of subjects, in a way that is entertaining and easy to absorb.

Each book contains 101 short accounts of key ideas or terms which are regarded as central to that subject. The accounts are presented in alphabetical order for ease of reference. All of the books in the series are written in order to be meaningful whether or not you have previous knowledge of the subject. They will be useful to you whether you are a general reader, are on a pre-university course, or have just started at university.

We have designed the series to be a combination of a text book and a dictionary. We felt that many text books are too long for easy reference, while the entries in dictionaries are often too short to provide sufficient

detail. The **Teach Yourself 101 Key Ideas** series gives the best of both worlds! Here are books that you do not have to read cover to cover, or in any set order. Dip into them when you need to know the meaning of a term, and you will find a short, but comprehensive account which will be of real help with those essays and assignments. The terms are described in a straightforward way with a careful selection of academic words thrown in for good measure!

So if you need a quick and inexpensive introduction to a subject, **Teach Yourself 101 Key Ideas** is for you. And incidentally, if you have any suggestions about this book or the series, do let us know. It would be great to hear from you.

Best wishes with your studies!

Paul Oliver
Series Editor

Acid Rain

Rain that has become unnaturally acidic, largely due to the presence of industrial pollutants in the atmosphere, is known as acid rain. Pure water is neither acidic nor alkaline. On the pH scale used to measure acid and alkali concentration water has a 'neutral' value of 7.0. All rainwater, however, is naturally slightly acidic. This is because carbon dioxide gas from the atmosphere and chloride from the sea dissolves in it to form carbonic acid (H_2CO_3) and hydrochloric acid (HCl). This gives unpolluted rainwater a pH value of around 5.6, lower values in the pH scale being associated with increasing acidity. Before the industrial revolution the pH of rain was generally between 5 and 6, in both urban and rural areas. Nowadays 'rain' with pH values substantially lower than 5 falls over vast areas of the world matching the generally accepted definition of 'acid rain'.

The main causes of acid rain are sulphur dioxide (SO_2) and nitrogen oxides such as NO and NO_2 (labelled collectively as NO_x). These are released during the burning of fossil fuels such as coal, natural gas, petrol, diesel and kerosene. The SO_2 is converted into sulphuric acid (H_2SO_4) on reaction with water, while the NO_x produce nitric acid (HNO_3).

Acid rain can damage living things and disturb natural ecosystems. It can kill fish, water plants and algae. It can wreak havoc on sensitive tree populations, particularly conifers. It also poses a direct threat to people by releasing toxic metals that would otherwise be retained within chemicals in the ground. Buildings, especially those made of limestone or marble, can be steadily eroded by reaction with acid rain. This has damaged many famous buildings worldwide.

The main solution to the acid rain problem is to reduce the release of SO_2 and NO_x from power stations and vehicles. Another option is to increase our use of alternative energy sources that do not involve the burning of fossil fuels.

see also...
Acids, Bases, pH

Acids

The digestion of food into nutrients our bodies can absorb and use, the pain of indigestion, the sharp taste of vinegar, starting a car, the death of fish in polluted lakes – these all involve the chemicals known as acids. Common acids are hydrochloric acid (HCl) which assists digestion of food in the stomach, sulphuric acid (H_2SO_4) found in car batteries, nitric acid (HNO_3) a component of acid rain, phosphoric acid (H_3PO_4) found in many foods and ethanoic acid (CH_3COOH) the active ingredient of vinegar.

The formula of each acid shown above includes hydrogen (H), because acids are commonly defined as chemicals that can release hydrogen ions (H^+). This is known as the 'Bronsted-Lowry' definition of an acid. Acids are the chemical opposites of bases, which accept hydrogen ions. Acids will react with bases in 'neutralization' reactions, to produce a 'salt' plus water. The salt is formed when the hydrogen ion released by the acid is replaced with a positive ion derived from the base. For example hydrochloric acid (HCl) will react with sodium hydroxide (NaOH) to form the salt sodium chloride (NaCl) and water (H_2O).

Acids are defined as strong or weak depending on the extent to which they dissociate into H^+ ions and the corresponding negative ions when they are dissolved in water. A strong acid, such as hydrochloric acid, dissociates completely:

$$HCl \rightarrow H^+ + Cl^-$$

A weak acid, like ethanoic acid, dissociates only partially, so the H^+ ions can recombine to form the acid molecule, a fact represented using a double-headed 'equilibrium' arrow:

$$CH_3COOH \rightleftharpoons CH_3COO^- + H^+$$

The concentration of the hydrogen ions released by acids is measured using the pH scale. Acidic pH values are those below 7.0. A reasonably concentrated solution of a strong acid will have a pH value of 1 or less. A weak acid at a similar concentration might have a pH around 4.

see also...

Acid Rain, Bases, Equilibrium, pH

Addition Reactions

In organic chemistry, an addition reaction is one in which two chemical groups add to the atoms linked by a double bond (one group to each atom), converting it to a single bond in the process. Some of the most important reactions of the chemical industry involve addition to the double bond in the simple alkene known as ethene. For example, the addition of water to ethene is used for the industrial manufacture of ethanol: A hydroxyl group (OH) from the water adds to one carbon atom of ethene, while a hydrogen atom (H) adds to the other carbon atom.

The sequential addition of identical molecules to form long chains produces many industrial polymers and plastics. Addition reactions among ethene molecules produce polyethene (polythene):

Replacing one of the hydrogen atoms of ethene with a chlorine atom allows the formation of polyvinyl chloride (PVC). Replacing all the hydrogen atoms of ethene with fluorine atoms yields polytetrafluoroethene (PTFE), known as 'Teflon' non-stick coating:

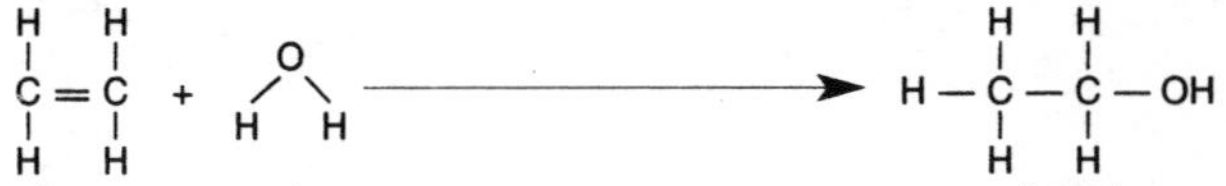

ethene   water   ethanol

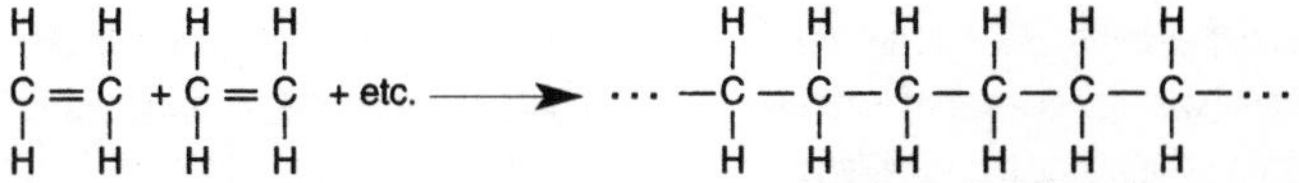

polyethene (polythene)

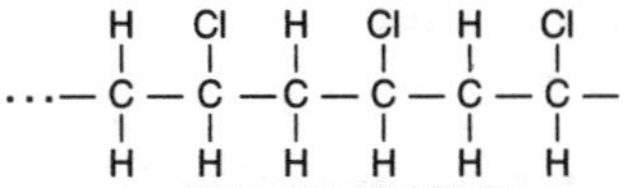

polyvinylchloride (PVC) 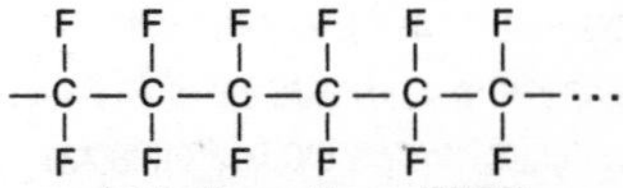   polytetrafluoroethene (PTFE)

see also...

Organic Chemistry, Polymers

Adenosine Triphosphate (ATP)

Life is an energy-requiring process, and many of the chemical reactions of life would not proceed if they were not coupled to other reactions that release energy. Within living cells, the chemical adenosine triphosphate (ATP) acts as a universal source of the chemical energy needed to make the chemistry of life happen.

This releases some of the tension that was in the ATP molecule, so it is accompanied by a release of energy. The process is known as the 'hydrolysis' of ATP, which literally means the breaking of ATP by reaction with water. This is the basic energy-releasing process that can be used to make other, energy-requiring, chemical reactions happen.

The structure of ATP includes a series of slight positive ($\delta+$) charges on phosphorus atoms arrayed near one another. These positive charges repel one another due to the 'like repels like' nature of the electric force. This, along with other aspects of ATP's structure, allows the ATP molecule to store useable energy, a bit like a compressed spring. When ATP reacts with water, a phosphate (PO_4^{3-}) group is removed from the molecule, to form adenosine diphosphate (ADP).

The chemical details of how the breakdown of ATP can power other reactions are quite complex. In general, however, energy-requiring reactions are *coupled* to the hydrolysis of ATP.

see also...

Condensation and Hydrolysis Reactions, Energy, Photosynthesis, Polar Covalency

Alcohols

In everyday life the word alcohol usually refers to the compound called 'ethanol', the active ingredient of alcoholic drinks. Both the pleasure and the problems caused by ethanol result from the way in which ethanol disturbs the normal processes of body chemistry. Ethanol is just one example of a vast range of alcohols – chemicals which contain at least one hydroxyl group (–OH) within their structure. Other common examples are isopropanol, the so-called 'rubbing alcohol' used in sterilising swabs; and glycerol, which is a 'tri-alcohol' or 'triol' because it contains three –OH groups. Cholesterol is a well-known and medically important complex alcohol. Chemists view alcohols as a broad category of chemicals with many useful properties, both as end-products, and as raw materials for

$$
\begin{array}{c}
\text{H} \\
| \\
\text{H} - \text{C} - \text{OH} \\
| \\
\text{H} - \text{C} - \text{OH} \\
| \\
\text{H} - \text{C} - \text{OH} \\
| \\
\text{H}
\end{array}
$$

glycerol
(propane–1, 2, 3–triol)

$$
\begin{array}{c}
\text{H} \\
| \\
\text{H} - \text{C} - \text{H} \\
\text{H} \quad | \quad \text{H} \\
| \quad | \quad | \\
\text{H} - \text{C} - \text{C} - \text{C} - \text{H} \\
| \quad | \quad | \\
\text{H} \quad \text{OH} \quad \text{H}
\end{array}
$$

2–methylpropan–2–ol

making other chemicals. They classify alcohols as 'primary', 'secondary' or 'tertiary' depending on the number of carbon atoms that are bonded to the carbon atom that carries the –OH group. A primary alcohol has no more than one carbon atom bonded to that carbon (e.g. ethanol); a secondary alcohol has two (e.g. propan–2–ol); while a tertiary alcohol has three (e.g. 2–methyl propan–2–ol).

$$
\begin{array}{c}
\text{H} \quad \text{H} \\
| \quad | \\
\text{H} - \text{C} - \text{C} - \text{OH} \\
| \quad | \\
\text{H} \quad \text{H}
\end{array}
$$

ethanol

$$
\begin{array}{c}
\text{H} \quad \text{H} \quad \text{H} \\
| \quad | \quad | \\
\text{H} - \text{C} - \text{C} - \text{C} - \text{H} \\
| \quad | \quad | \\
\text{H} \quad \text{OH} \quad \text{H}
\end{array}
$$

isopropanol
(propan–2–ol)

see also...
Functional Groups

Alkali Metals

The metals in Group 1A of the Periodic Table, namely lithium (Li), sodium (Na), potassium (K), rubidium (Rb), caesium (Cs) and francium (Fr), are known as the alkali metals. These metals all react vigorously with water to form alkali solutions. Sodium, for example, forms sodium hydroxide (NaOH) solution.

$$2Na \text{ (s)} + 2H_2O \text{ (l)} \rightarrow 2NaOH \text{ (aq)} + H_2 \text{ (g)}$$

The common name for sodium hydroxide is 'caustic soda', traditionally used to clear blocked drains because it reacts with and disperses fats and other animal and vegetable wastes.

When they react, the atoms of alkali metals each lose their outer electron to form singly charged positive ions, such as Na^+. This means the alkali metals are powerful 'reducing agents' – agents that deliver electrons to other chemicals when they react. The alkali metals are so reactive they occur naturally on Earth only combined within compounds. If we prepare a sample of a pure alkali metal, it will begin to react immediately with oxygen unless we take steps to prevent this from happening. This explains why sodium, for example, is stored in laboratories under unreactive oil.

Sodium chloride (NaCl), common 'table salt', is a familiar example of a compound of an alkali metal, and is also the major source of sodium in our diet. Ions of sodium and potassium are dietary essentials. They are crucial 'electrolytes' – charged particles dissolved in body fluids, which play vital roles in the working of all cells, especially nerve cells. Too much sodium in the diet, however, has been linked with high blood pressure and other problems. Increasing our dietary intake of potassium, on the other hand, is a recommended measure to reduce blood pressure.

The only economic way to produce pure alkali metals is to release them from their compounds by electrolysis. This uses an electric current to force electrons onto the metal ions. For example: $Na^+ + e^- \rightarrow Na$.

see also...

Electrolysis, Metals, Oxidation, Reduction

Antibiotics

Any substance that can inhibit the growth of or destroy micro-organisms can be described as an antibiotic. Most antibiotics are active against bacteria, indeed some people use the term to refer to agents that are effective only against bacteria. Antibiotics do not affect viruses. The best known antibiotic is probably penicillin, a product of fungi which inhibits the growth of bacteria by interfering with the manufacture of their cell walls. The story of Alexander Fleming discovering penicillin in 1928 is standard school science. This was certainly a key step in the mass application of antibiotics, but antibiotics have been known of and used to treat illness for over 2500 years. The ancient Chinese probably discovered them first, when they found that mouldy soya bean curd could be used to treat infections.

Modern medicine makes use of thousands of antibiotics. These include natural products like penicillin, synthetic compounds and 'semi synthetic' antibiotics created by chemically modifying natural antibiotics. The antibiotics are grouped according to their chemical structures under such familiar categories as the penicillins, cephalosporins, erythromycins and tetracyclines.

Many antibiotics work by acting as enzyme inhibitors: chemicals that can bind to specific sites on the surface of enzymes and inhibit the enzymes from participating in the acts of catalysis that sustain the target organisms' biochemistry. Penicillin, for example, binds to and inhibits an enzyme that catalyses a crucial step in the manufacture of bacterial cell walls. Some antibiotics are also useful in the treatment of cancer, leading to their classification as 'anti-tumor antibiotics'.

Antibiotics have transformed the treatment of infections, but their success is threatened by the growing problem of antibiotic resistance. This is the emergence of strains of bacteria that are resistant to the most commonly used antibiotics.

see also...
Enzymes

Antibodies

The body is protected against infectious disease by the immune system, sustained by specialized immune cells that circulate in blood and lymph. Some of the most powerful chemicals of the immune system are the antibodies that are produced in response to specific threats. An antibody is a protein composed of several distinct protein molecules bonded together by 'disulphide' bonds between sulphur atoms (–S–S–). Crucial parts of the antibodies have 'variable' structures which are able to bind to parts of foreign molecules defined as 'antigens'. As soon as an antigen enters the body it stimulates the production of antibodies that can bind to it.

Antibodies help the immune system to eliminate invading micro-organisms in various ways. By binding to the surface of a virus, for example, an antibody may neutralize the virus and prevent it from infecting our cells. Antibodies help cells of the immune system to engulf and digest foreign material. The binding of an antibody to an antigen also activates a series of blood-borne enzymes known as 'complement', which can disrupt and destroy invading micro-organisms.

Once we have caught and successfully fought off any particular infection, antibodies able to bind to the associated antigens will remain circulating in our blood. These circulating antibodies are part of the defensive process that can leave us 'immunized' against subsequent attack by the same infection. The antibodies in our blood also provide evidence of our exposure to a particular infectious agent. The process of intentional immunization known as vaccination involves stimulating an immune response, including production of antibodies, by administering an antigen in an attenuated form. For example dead cells of a particular micro-organism, or purified antigens from these cells, can be used as a vaccine. Antibodies can also be purified from blood and administered to help another person fight off an attack of the appropriate infection. This procedure is known as 'passive immunization'.

see also...
Enzymes, Proteins

Aromatic Compounds

To a chemist, an aromatic compound is one that contains at least one benzene ring, so the simplest aromatic compound is benzene itself (C_6H_6), shown below:

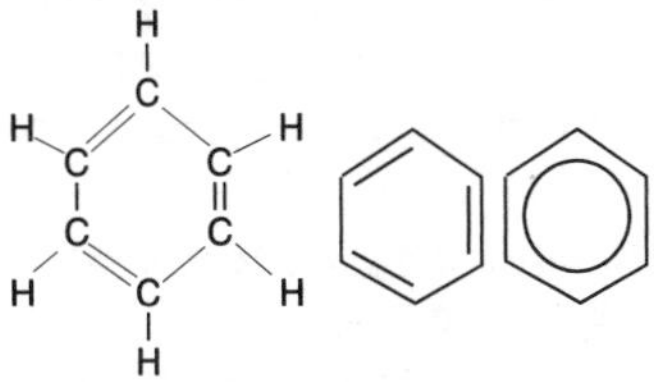

The three representations of benzene, C_6H_6

The structure on the left is the simplest way to represent benzene using conventional single and double bonds. The central structure shows the same thing using a common chemical shorthand for organic chemicals, in which carbon atoms are at the vertices where bonds meet, and hydrogen atoms are not shown. We know that these views of benzene are not strictly accurate, however, because the six carbon-to-carbon bonds in the benzene ring are actually identical. This is caused by an important chemical phenomenon called *electron delocalization*. The six electrons that are needed to form the three second bonds of each double

bond in the structure on the left are actually free to roam around the six-carbon ring. In other words, they are 'delocalized', with each carbon atom, on average, carrying one of the six delocalized electrons. Electron delocalization can occur whenever a chemical appears to contain alternating double bonds. The structure on the right hand side is, therefore, a more accurate representation of benzene, with a circle representing the ring of delocalized electrons.

Aromatic compounds got their name because some of the first to be identified had powerful aromas, but the presence of at least one benzene ring is their distinguishing feature, regardless of any odour they may have. Aromatic compounds are widespread in nature, and form some of the most important chemicals of life. They are also widely used by synthetic chemists to make such things as plastics, pigments and drugs.

see also...

Organic Chemistry

Atoms

The fundamental building blocks of all chemicals are the tiny particles of matter known as atoms. Just over 100 types of atoms are known. Any substance composed of only one type of atom is called an *element*, so all the atoms are listed in the Periodic Table of the Elements (page 103).

Atoms are composed of much smaller 'sub-atomic particles', known as protons, neutrons and electrons. Protons have a positive electric charge of +1 while electrons have a negative electric charge of –1. Each atom is electrically neutral overall, since it contains an equal number of protons and electrons. The protons and neutrons are bound together within the dense central 'nucleus' of an atom. The electrons move around the nucleus in regions of space known as electron orbitals. Familiar, but over-simplified depictions of atoms show the electrons orbiting the nucleus like satellites.

The simplest atom, hydrogen, contains just one proton and one electron. The largest naturally occurring atom, uranium, contains 92 protons and 92 electrons. The number of neutrons within atoms varies. Atoms of the same element that differ in the number of neutrons they contain are known as 'isotopes' of the element.

Each atom is characterised by two numbers: the 'atomic number' is the number of protons, the 'mass number' is the number of protons plus the number of neutrons. This information, together with the symbol for each atom, is often presented in 'nuclide notation', as follows:

mass number

$$^{12}_{\ 6}\text{C}$$

atomic number   symbol for carbon
Nuclide notation for carbon – 12

A simple depiction of atomic structure, with electrons 'orbiting' the nucleus

nucleus, composed of protons and neutrons

electron

see also...

Electron Orbitals, Elements, Isotopes, Periodic Table, Sub-atomic Particles

Bases

The chemical opposites of acids are known as bases. A base can be defined as any chemical that can combine with a hydrogen ion (H^+). This is known as the 'Bronsted-Lowry' definition of a base. Any solution of a base in water is called an 'alkali'. Common bases are sodium hydroxide ($NaOH$) known as 'caustic soda', calcium oxide (CaO) known as 'quicklime', calcium hydroxide ($Ca(OH)_2$) known as 'slaked lime', potassium hydroxide (KOH) known as 'caustic potash', and ammonia (NH_3).

Bases react with acids in 'neutralization' reactions to produce a 'salt' plus water. The salt is formed when a positive ion derived from the base combines with a negative ion derived from the acid. For example, the base potassium hydroxide (KOH) will react with nitric acid (HNO_3) to form the salt potassium nitrate (KNO_3) which is used as a fertilizer.

The soluble bases we call alkalis are defined as strong or weak, depending on the extent to which they dissociate or react to form OH^- ions and the corresponding positive ion when they interact with water.

A strong alkali, such as sodium hydroxide, dissociates completely:
$$NaOH \rightarrow Na^+ + OH^-$$
A weak alkali, like ammonia, sets up an equilibrium situation in which much of the alkali remains in its original form, so causing fewer than the maximum number of OH^- ions to be generated:
$$NH_3 + H_2O \rightleftharpoons NH_4^+ + OH^-$$
Alkalis have pH values greater than 7, with the concentration of OH^- ions increasing as the pH value rises.

Bases, including alkalis, can be highly corrosive substances just like their opposites, the acids.

In recent years, one particular class of bases has attained a high media profile, namely the four nitrogen-containing bases of DNA. Genetic differences between different people, and between humans and other species, are determined by the sequence in which these bases (adenine, thymine, guanine and cytosine) are arranged in DNA.

see also...

Acids, Deoxyribonucleic Acid (DNA), Equilibrium, pH

Batteries

A battery is a device which employs a chemical reaction to generate an electric current. The system at the heart of all batteries is an 'electrochemical cell'. Strictly speaking, a battery is a collection of electrochemical cells working together, but even a single electrochemical cell is commonly referred to as a battery.

The chemicals within a battery participate in a 'redox' (reduction-oxidation) reaction, meaning a reaction which involves a flow of electrons from some chemicals to others. The design of the battery ensures that the electrons are generated at one end or terminal of the battery, and need to flow into the other end or terminal. So when the two terminals are connected by an external circuit, the electrons flow through the circuit and do useful work such as lighting a torch bulb.

In a typical torch battery, the source of the electrons is the oxidation of zinc into zinc ions:

$$Zn \rightarrow Zn^{2+} + 2e^- \text{ (oxidation)}$$

When these electrons flow back into the battery at the other terminal they combine with manganese dioxide (MnO_2) to form negative MnO_2^- ions.

$$MnO_2 + e^- \rightarrow MnO_2^- \text{ (reduction)}$$

So the reaction of zinc with MnO_2 is the source of electrical power.

By ensuring the electrons can only be transferred from Zn to MnO_2 through an external circuit, we can make the reaction generate an electric current, rather than having all the energy wasted as heat. Additional reactions besides the one shown here do occur, but we have shown the principal reactions generating the electric current.

In a rechargeable battery, the chemical reactions that produce an electric current can be reversed if a current is applied to the battery instead. Car batteries are important examples of rechargeable batteries. They use chemistry to provide the spark needed to start the engine or power lights, etc. when the engine is off. When the engine is running it drives a generator (the alternator) that produces the electricity needed to recharge the battery.

see also...

Oxidation, Reduction

Biochemistry

All living things, including humans, are built of chemicals and sustained by chemical reactions. The chemistry of life is called biochemistry. It must be powered by a flow of energy, and the ultimate source of that energy is sunlight. Energy from the sun powers photosynthesis in plants, which converts water and carbon dioxide into carbohydrates and releases oxygen gas. The eventual oxidation of these carbohydrates back to water and carbon dioxide then releases the energy needed to power every other chemical reaction of life. Animals, including humans, are sustained by eating plants directly, or by eating animals that are themselves sustained by eating plants. So all biochemistry is 'solar powered'.

The many thousands of chemical reactions in living things are catalysed by protein molecules called *enzymes*. Without enzymes there would be no life. Different organisms contain different enzymes and other proteins. It is this that is responsible for the variety of life forms. Protein composition, in turn, is determined by the DNA sequence of an organism's genes. Genes embody a chemical code that determines the sequence in which amino acid molecules are linked together to form proteins. The proteins then 'do the work' of actually constructing and maintaining life, by serving as enzymes, structural proteins, and performing many other roles.

The chemistry of life is known as the process of 'metabolism'. Metabolism is itself split into two branches: 'catabolism' in which chemicals such as foods are *broken down* to release energy and raw materials; and 'anabolism', in which raw materials are *built up* into the chemicals of cells and bodies.

Since biochemistry is the chemistry of life, it is also the chemistry of illness, infection and medicine. Every illness has some biochemical cause or consequence, often both. Medicines are chemicals used to interfere with biochemical processes.

see also...

Carbohydrates, Deoxyribonucleic Acid (DNA), Enzymes, Fats and Oils, Photosynthesis, Proteins, Ribonucleic Acid (RNA)

Calorimetry

Looking at food labels to find out how many 'calories' a food supplies is an everyday event for those of us trying to maintain an appropriate energy intake. Food releases energy when the chemicals within it combine with oxygen, a process that is identical, overall, to burning or 'combustion'. So, we can determine the maximum amount of energy a food can release to the body by burning a sample completely in oxygen, and measuring the amount of heat energy released. This procedure is known as calorimetry.

Calorimetry takes its name from the unit of energy called the 'calorie', but the standard unit of energy is now the *joule* (J). One calorie equals 4.184 joules. Unfortunately the terminology is confused by the use of Calorie (with a capital C) to indicate 1000 calories (1 kilocalorie or kcal). So when we talk of a banana supplying us with '70 Calories', we really mean 70,000 calories which is 70 kilocalories (70 kcal).

The apparatus used to determine the energy content of a sample of food is known as a calorimeter. There are many different types of calorimeter but typically they have a central chamber containing the material to be burned and an excess of oxygen. The material is ignited (by an electrical spark, for example) and the heat energy is released to flow into a heat reservoir, such as a known mass of water or metal surrounded by an effective insulating layer. The rise in temperature is measured. This allows the amount of heat energy released to be calculated, for example, using the equation:

$$Eh = m \times c \times \Delta T$$

where Eh = heat energy in joules, c = the known specific heat capacity of the reservoir (i.e. how many joules are required to raise the temperature of 1 gram by 1 degree Celsius), m = the mass of the reservoir in grams, and ΔT = the temperature rise in degrees Celsius.

This is the basic procedure that has been used to determine, for example, that fats and oils in food supply considerably more energy than carbohydrate or protein.

see also...

Carbohydrates, Energy, Fats and Oils, Proteins

Carbohydrates

lucose (blood sugar), sucrose (table sugar), starch and cellulose are all examples of the chemical compounds known as carbohydrates. Their name indicates that they contain carbon and the elements found in water, a link further established by their general formula of $C_x(H_2O)_y$. The simplest carbohydrates are the monosaccharides (meaning 'single sugars') such as glucose. Disaccharides ('double sugars'), such as sucrose, contain two monosaccharide units bonded together.

Glucose

The other major category of carbohydrates is the polysaccharides ('many sugars') which can contain many thousands of monosaccharide units bonded into straight or branched chains. Starch and cellulose are polysaccharides composed of many glucose units.

Carbohydrates are major constituents of living things, in which they serve as a source of chemical energy. The oxidation of carbohydrates to carbon dioxide and water releases around four kilocalories of energy per gram of carbohydrate. The carbohydrates are a major source of energy in our diet. Plants generate carbohydrates from carbon dioxide and water in photosynthesis, powered by sunlight. Life on Earth is sustained by the continual cycling of carbon, hydrogen and oxygen through photosynthesis, which makes carbohydrates, and cellular respiration, which uses the carbohydrates as a source of energy.

Some carbohydrates become bonded to protein molecules to form 'glycoproteins'. These are vital for intercellular recognition processes, and in many of the chemical interactions involved in the immune system and the onset of disease. The study of these roles of carbohydrates is known as 'glycobiology', and is one of the most active fields of modern medical research.

see also...

Calorimetry, Energy, Photosynthesis

Carbon Cycle

The element carbon travels through different chemical forms on planet Earth in a network of chemical processes called the carbon cycle. The most dynamic part of the carbon cycle is the continual cycling of carbon between the carbon dioxide in the atmosphere and the many carbon-containing compounds in living things. Carbon dioxide is incorporated into the organic chemicals of plants during photosynthesis, then into animals when they eat plants, then released back to the atmosphere when food is combined with oxygen during respiration.

A much larger but slower part of the carbon cycle involves carbon in the sea and the land. Huge quantities of carbon are stored in 'sedimentary rocks', such as limestone, chalk and marble – types of calcium carbonate ($CaCO_3$). These sedimentary rocks have formed over ages from carbon in the oceans, in the form of dissolved carbon dioxide (CO_2), hydrogen carbonate ions (HCO_3^-), carbonate ions (CO_3^{2-}) and carbonic acid (H_2CO_3). Sedimentary rocks also contain carbon derived from calcium carbonate in skeletons and shells of marine organisms. When geologic activity raises the sedimentary rocks above the surface they gradually erode and return their carbon to the seas. In this way carbon moves through the seas and the land in a relentless geologic cycle.

Other geologic reservoirs of carbon are the 'fossil fuels' – coal, oil and natural gas – derived from ancient plant and animal life. In recent years human activity has significantly disturbed the natural carbon cycle by burning, in a very short period of time, huge amounts of fossil fuel that had accumulated over millennia. This has released into the atmosphere the carbon dioxide that is believed to be a major factor in global warming. While burning so much fossil fuel, we have also been cutting down and burning the rainforests that play a significant role in mopping up carbon dioxide in the atmosphere and storing it in new plant growth.

see also...

Global Greenhouse Effect, Nitrogen Cycle, Organic Chemistry, Photosynthesis

Carbon Dating

As soon as a living organism dies a 'clock' starts ticking within its chemicals that can be used to estimate the time of death on a timescale of thousands of years. The age of a piece of wood, the leather of a shoe, the cloth of a natural fabric, or any other material derived from animals or plants is revealed by the ratio of two isotopes of carbon within the sample. Living things have a tiny, known, proportion of carbon-14 within them, which is believed to have remained essentially constant for thousands of years. Carbon-14 is a radioactive isotope that decays into non-radioactive products, so the ratio of carbon-14 to the other natural isotopes of carbon (carbon-12 and carbon-13) begins to fall as soon as an organism dies. The amount of carbon-14 remaining is related to the intensity of the radiation produced by its decay. The half-life of carbon-14 is 5700 years, so the radiation due to carbon-14 falls to half the original value after 5700 years, to one quarter after 11400 years (i.e. 2×5700), and so on. The radiation from the decay of carbon-14 can be detected by sensitive radiation monitors and compared with modern material similar to the sample being examined. A simple mathematical formula yields an estimate of the sample's date.

This technique has been used to 'carbon-date' archaeological samples as far back as 70 000 years. The most famous example was the analysis of the Turin Shroud, reputedly used to wrap the body of Christ. Carbon dating suggested that the Shroud was a mediaeval fake. The controversy resulting from this analysis publicized some of the difficulties and uncertainties in the technique. Contamination of an ancient sample with more recent material could invalidate a result. There is also debate about the validity of the assumptions made about carbon-14 levels thousands of years ago.

Despite the controversies and some uncertainties, carbon dating is generally regarded as a reliable method for dating ancient materials derived from living things.

see also...

Atoms, Isotopes, Radioactivity

Carcinogen

Any chemical that can cause the unregulated growth and multiplication of cells we call cancer is known as a carcinogen. The effect of many carcinogens is due to their ability to interact with DNA in a way that causes changes in DNA known as 'mutations'. Carcinogens that act in this way are also known as 'mutagenic carcinogens' or just 'mutagens'. The mutations are most likely to cause cancer when they occur in genes that code for crucial proteins involved in regulating cell growth and cell division. Some carcinogens do not appear to act directly on DNA, so they are not mutagens and must cause cancer in more complicated indirect ways that are not well understood.

Many mutagenic carcinogens have been found to behave chemically as 'electrophiles' (electron-lovers), meaning they can bind to electrons or negatively charged groups in DNA. Others, although not themselves electrophiles, are converted into electrophiles when they interact with the metabolic enzymes in a cell. Some carcinogens are 'free radicals', which are unstable chemicals with an unpaired electron. Free radicals tend to participate in reactions that yield an electron for the single electron of the free radical to pair up with, causing damage to the source of the extra electron, such as a molecule of DNA. When a chemical loses an electron (to a carcinogen, for example) we say it has been oxidized. Chemicals that can react with and neutralize carcinogens are often classified as 'antioxidants', because they counteract the oxidizing powers of the carcinogens.

Potent carcinogens occur in smoke, including, of course, tobacco smoke and some of the chemical effluents produced by industrial processes. Many known carcinogens, however, also occur in everyday foods such as the fruits and vegetables we are encouraged to eat to protect us against cancer. One crucial factor appears to be *dosage*. Routine exposure to low levels of carcinogens is inevitable, and can be endured for years without harm.

see also...

Deoxyribonucleic Acid (DNA),
Electrophiles and Nucleophiles,
Free Radicals, Oxidation

Catalyst

A catalyst is any chemical that speeds up a particular chemical reaction without itself being permanently altered in the process. A tiny quantity of catalyst can cause a big increase in reaction rate. Catalysts work by lowering the activation energy for the reactions they catalyse. The presence of a catalyst increases the proportion of reactant molecules that have sufficient energy to react. Catalysts can be very specific, each one catalysing only a particular reaction, or a limited range of reactions. The chemical industry makes great use of catalysts to accelerate the reactions involved in making the materials, fabrics and medicines that sustain our modern way of life. Living things rely on biological catalysts called enzymes to catalyse almost all of the chemical reactions of life.

The 'catalytic converters' in car exhausts are made of honeycomb gauzes coated with platinum, palladium and rhodium. These elements catalyse the conversion of polluting gases such as NO and CO into less hazardous forms. For example, platinum catalyses the conversion of poisonous carbon monoxide into carbon dioxide. The 'biological detergents' many of us use to clean our clothes contain enzymes that catalyse the break down of proteins and fats into simpler substances which can be readily rinsed away.

A catalyst that is in the same physical phase as the chemicals whose reaction it catalyses is called a *homogeneous catalyst*. A catalyst in a different physical phase is called a *heterogeneous catalyst*. The platinum of a car's catalytic converter is a heterogeneous catalyst, since it is a solid catalysing a reaction involving gases. Examples of homogeneous catalysts are the enzymes dissolved in our intestinal fluid, catalysing the digestion of food molecules which are also dissolved in that fluid.

Catalysts can be inactivated by chemicals known as 'catalyst poisons', which can bind to them and alter their structure or properties in some way.

see also...

Chemical Reaction, Enzymes

Chain Reaction

A chain reaction is a reaction that releases one or more products which can initiate the same reaction in other molecules (or atoms or ions). This means that chain reactions are self-sustaining, and will proceed very quickly until all the reactants are used up or until some chemical is added to terminate the reaction. Many chain reactions are initiated by light energy. For example, a mixture of hydrogen and chlorine will begin to react when light energy breaks a chlorine molecule into two Cl atoms:

$$Cl_2 + \text{light energy} \rightarrow Cl\bullet + Cl\bullet$$

The Cl atoms are free radicals, i.e. each has an unpaired electron ($\bullet$), and so are highly reactive. Each one can react with hydrogen:

$$Cl\bullet + H_2 \rightarrow HCl + H\bullet$$

This generates a hydrogen free radical which can react with chlorine:

$$H\bullet + Cl_2 \rightarrow HCl + Cl\bullet$$

generating another Cl free radical which reacts with hydrogen as before:

$$Cl\bullet + H_2 \rightarrow HCl + H\bullet$$

Each step in the chain reaction generates a free radical that will initiate the next step.

The involvement of free radicals is typical of many chain reactions. The formation of many synthetic polymers is achieved by addition reactions that are initiated by an appropriate free radical, and can be terminated by adding a chemical that will combine with the free radicals.

One of the most publicized chain reactions, and one that does not involve free radicals, is the process used to amplify tiny samples of DNA. This is called the Polymerase Chain Reaction (PCR), because it relies on the enzyme DNA polymerase to make copies of DNA. The PCR system allows tiny samples of DNA obtained from blood stains or semen, for example, to be copied into a large number of identical copies suitable for 'DNA fingerprinting'. The use of PCR and DNA fingerprinting has determined the guilt or the innocence of many suspects in murder, rape and other crime cases.

see also...

Addition Reactions, Chemical Reaction, Deoxyribonucleic Acid (DNA), Free Radicals

Chemical Reaction

The fundamental concept involved in our understanding of chemical change is the chemical reaction. This is a process in which chemicals called 'reactants' encounter one another, and *react* to the encounter in a way that generates different chemicals called 'products'

REACTANTS → PRODUCTS

Each reaction is accompanied by some energy change in which energy will either be released from the chemicals to the surroundings, or absorbed into the chemicals from the surroundings. Reactions that release energy are 'exothermic' reactions, while those that absorb energy are 'endothermic reactions'.

Chemical reactions are initiated by *collisions* among the atoms, molecules or ions involved. The collisions occur because all particles are moving, i.e. have some kinetic energy, with the temperature of a chemical being related to how quickly its particles are moving. Our understanding of the role of collisions in chemical reactions is labelled 'collision theory'. Many collisions between particles simply result in the particles bouncing off one another

unchanged. In order to initiate a chemical reaction a collision must be sufficiently violent, or 'energetic' to disrupt the electronic structure of the chemicals enough to permit a chemical change. In all cases, the chemical change involves a rearrangement of electrons to form different chemical bonds from those in the reactants. A chemical reaction is a process of *electron rearrangement*.

The minimum energy a collision must involve in order to promote a reaction is called the 'activation energy' of the reaction. As temperature increases, more collisions will occur with an energy at least as great as the activation energy of the reaction. This explains why heating chemicals causes the rate of a chemical reaction to increase. It is one reason why chemical laboratories contain Bunsen burners, and other heating devices, which chemists use to jolt reluctant chemicals into reacting.

see also...

Covalent Bonds, Electron Configuration, Ionic Bonds

Chemotherapy

The use of chemicals to treat disease is known as chemotherapy. That very broad definition applies the term chemotherapy to every drug and medicine, covering many thousands of natural and human-made compounds and also several elements. 'Antimicrobial chemotherapy' is the use of chemicals to treat infectious diseases, caused by any of the three main categories of infectious agents – bacteria, viruses and fungi. Many of the chemicals used in antimicrobial chemotherapy are referred to as antibiotics, with familiar names such as penicillin and tetracycline. 'Anticancer chemotherapy' is the use of chemicals to treat cancer. This is the sense in which the term chemotherapy is most commonly used today.

To be an effective chemotherapy agent a chemical must be able to enter an appropriate part of the body, be transported (if necessary) to the site of disease, perhaps pass through cell membranes to enter cells, then have the desired therapeutic effect. The way in which a chemotherapy agent achieves all these depends on the precise chemical structure of the agent concerned. In some cases the chemotherapy agent may be delivered into the body by coating it with other chemicals, chosen to facilitate the crucial entry phase.

Anticancer agents generally work by killing or at least suppressing the growth of cancer cells. Some achieve this by reacting with DNA, while others bind to and inhibit enzyme molecules involved in cell growth and division. The key to successful use of such agents is to find ways to ensure they preferentially attack cancer cells, while leaving healthy cells relatively unharmed. Many chemotherapy agents are natural products found in organisms such as plants and fungi. Others are natural products that have been chemically modified to enhance their effects. Many are completely synthetic compounds devised by chemists.

see also...

Antibiotics

22

Chromatography

One of the most fundamental challenges facing chemists when they study and synthesise specific chemicals is to separate out the individual components of a mixture of chemicals. This may be necessary in order to identify the chemicals in a sample as part of a medical, environmental or forensic analysis. Or it may be needed to prepare a pure sample of a desired chemical, such as a drug or a chemical intermediate in the manufacture of another chemical product. Chromatography is one of the oldest and most versatile ways of separating the components of a chemical mixture.

There are many types of chromatography but all involve a mixture of chemicals carried in the 'mobile phase', moving across or through another chemical called the 'stationary phase'. The components of the mixture move at different rates, depending on the ease with which they mix with and are attracted to the stationary phase and the mobile phase. This causes the mixture to separate into bands of relatively pure components as they move through the chromatography system.

The name 'chromatography' is derived from the fact that the chemicals being separated are sometimes different colours (*chroma* means colours). For example, many of us have seen school laboratory experiments in which different coloured dyes are separated from black ink, when a solvent is used to move the ink through absorbent paper. In such 'paper chromatography', the paper forms the stationary phase and the solvent carrying the ink through the paper is the mobile phase. Another chemical commonly used as the stationary phase is silica (silicon dioxide – SiO_2), which can be packed into tubular columns or spread across glass plates. In 'high pressure liquid chromatography' (hplc), liquid mobile phases are pushed at high pressure through columns packed with an appropriate solid phase. Gases can also be used as the mobile phase in 'gas chromatography' (gc).

The more sophisticated and automated methods of chromatography are some of the most precise tools available to analytical chemists.

Combinatorial Chemistry

Chemists have traditionally found useful compounds either by the chance discovery of interesting properties in a particular compound, or by laboriously trying to synthesize specific 'target' compounds which they hope will have the properties they desire.

In recent years a new technique has become increasingly popular which actually mimics important aspects of the way in which evolution has crafted the chemistry of life. This new approach, called combinatorial chemistry, involves using automated or semi-automated procedures to generate large numbers of compounds by bringing together simpler starting materials in a huge variety of possible combinations. The resulting large pool of compounds can then be screened in some efficient and possibly automated way to identify specific compounds with chosen properties. An example would be the ability to interact with a particular biochemical molecule in a way that might make the compound a useful drug. This is similar to the process of evolution by natural selection, in which random genetic changes create biochemical novelties, including new chemical compounds. The difference is that in combinatorial chemistry the selection of the new compounds is performed deliberately by chemists and their screening procedures, rather than by the automatic 'survival of the fittest' process of natural selection.

Combinatorial chemistry is now routinely used in the pharmaceutical industry by chemists seeking new drugs. Once a promising candidate emerges, however, more traditional skills can be used to modify the candidate compound step-by-step, until the most effective compound or range of compounds is found. Combinatorial chemistry mixes the best effects of chance and conscious design to arrive at a desired result more quickly and efficiently than either chance or design would achieve on their own.

Having proved successful in the pharmaceutical industry, combinatorial chemistry is increasingly being applied to other tasks, such as the search for new materials with interesting electrical, magnetic, optical or catalytic properties.

Compounds

Whenever two or more elements are bonded together in fixed proportions a chemical compound is formed. A virtually limitless number of compounds exist, all made from differing combinations of the approximately 100 elements listed in the Periodic Table. Two very simple compounds are sodium chloride (NaCl) and water (H_2O). The formulas indicate the proportions in which the elements involved are present. These examples illustrate the principle that the chemical properties of a compound are very different from the properties of the individual elements in the compound. Sodium chloride is common salt, a solid compound that is essential for life and which we happily add to our food. Yet it is formed from sodium, a metal that reacts explosively with water, and chlorine, a poisonous gas. Similarly, water, a liquid essential for life and which can be used to put out fires, is formed from hydrogen, an explosive gas, and oxygen, the gas needed to sustain burning. Compounds range from simple structures such as hydrogen chloride (HCl), containing just two elements bonded into tiny diatomic molecules, up to giant structures composed of huge numbers of many different kinds of atoms.

Compounds can be classified into covalent compounds (such as water) held together by covalent bonds, and ionic compounds (such as sodium chloride) held together by ionic bonds. They are also classified as either organic compounds, with a framework of bonded carbon atoms; or inorganic compounds, built largely from elements other than carbon. Organic compounds deserve their special designation because they form the chemical basis of life and of most medicines, fabrics and plastics.

Each substance is either a compound, an element or a mixture. The crucial distinction between compounds and mixtures is that the parts of a mixture are just loosely gathered together, and are readily separated by physical means. The component elements in compounds can only be separated by a chemical reaction.

see also...

Covalent Bonds, Elements, Formula, Ionic Bonds, Organic Chemistry, Periodic Table

Computational Chemistry

Computers and computer science are nowadays applied to chemical challenges in ways that are sufficiently distinctive to be known as the field of 'computational chemistry'. Chemistry has traditionally progressed by *experimental chemistry*, in which chemicals are manipulated and studied; and *theoretical chemistry*, based on reasoning and calculations. Computational chemistry has been called 'a third way of doing chemistry', using computers to perform complicated calculations or simulations which would not be feasible by any other means.

A good example of computational chemistry in action comes from the search for new drugs. Many research groups are tackling this using sophisticated computer simulations to identify molecules that might be able to bind to enzymes and other biomolecules involved in the development of cancer. The structures of millions of molecules can be examined in these simulations, allowing promising candidates to be examined more closely in real experiments. There are already drugs on the market that were first identified as possible therapeutic agents by this kind of 'virtual testing'.

A related example is work to predict the way in which new or modified protein molecules might fold up into the complex 3-dimensional structures that allow them to perform their functions. The folding process is notoriously complicated, but computer simulations are growing ever more accurate at predicting the course of this crucial biochemical process. One aim is to help chemists to create modified or synthetic proteins that can be used as medicines. Another aim is simply to understand the chemistry of life in more detail. Computational chemistry is also powering advances in theoretical chemistry, such as our understanding of the shapes and properties of the molecular orbitals of large molecules.

see also...

Electron Orbitals, Proteins, Quantum Chemistry

Condensation and Hydrolysis Reactions

In the chemistry of life, when small molecules need to be linked up into large molecules Nature generally relies on condensation reactions. They get their name from the fact that water molecules are a by-product, with one water molecule being formed from a hydrogen atom (H) derived from one of the reactants and a hydroxyl group (OH) from the other. Proteins, for example, are formed when condensation reactions link many amino acids together. This is represented by the top arrow in the diagram below (which shows two amino acids becoming linked). The reversal of a condensation reaction is a hydrolysis reaction – the breakdown of a large molecule by reaction with water. This is represented by the lower reaction arrow.

DNA, RNA and complex carbohydrates such as starch, cellulose and glycogen are all examples of large 'condensation polymers' made when smaller monomer units become linked by condensation reactions. Esters and polyesters are also formed in condensation reactions. Another vital condensation and hydrolysis process is the formation of the energy storage molecule ATP, in a condensation reaction; and its breakdown to ADP and phosphate in a hydrolysis reaction when the energy stored in ATP is released to power other chemical reactions.

The term condensation reaction also covers reactions in which the small molecule released is not water, but may be HCl, for example.

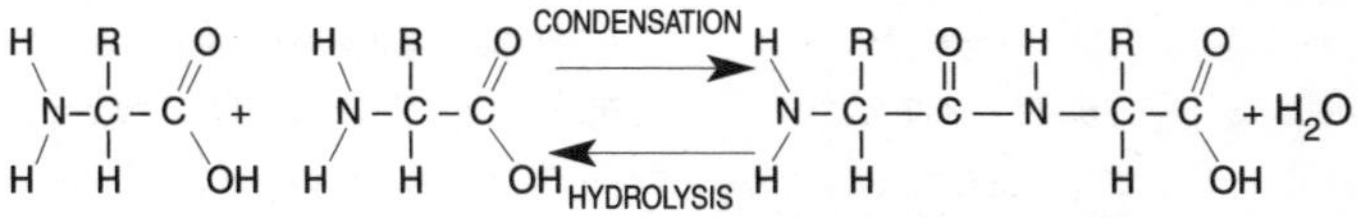

When proteins need to be broken down, during the digestion of food for example, this proceeds by way of hydrolysis reactions breaking the bonds between the protein's amino acids.

see also...

Adensosine Triphosphate (ATP),
Deoxyribonucleic Acid (DNA),
Esters, Polymers, Proteins,
Ribonucleic Acid (RNA)

Coordination Complexes

When a positively charged metal ion becomes surrounded by a fixed array of negative ions or molecules that have free electron-pairs able to interact with the metal ion, a coordination complex (or 'coordination compound') can be formed. The molecules or ions bound to the central metal ion are called 'ligands'. If a ligand has two electrons interacting with the metal ion, the electron-pair is said to form a 'coordinate bond'. The number of ligand sites bound to the metal ion is the 'coordination number' of the complex. Some complex ligands bind to the metal ion at more than one site. For example, both ends of certain U-shaped ligands may be coordinated to the central ion, making them 'bidentate ligands' (bidentate because they take 'two bites' at the ion). Other complex ligands can be tridentate, tetradentate, and so on.

Many important chemicals in living things are coordination complexes. Some enzymes and other proteins include metal ions bound in the form of coordination complexes at their 'active sites' – the sites responsible for their characteristic chemical activities. The protein haemoglobin, for example, includes a coordination complex of an iron ion bound to nitrogen atoms that are part of the 'haem' group bound to the protein chain. One binding site remains free, available to temporarily bind to oxygen molecules and allow haemoglobin to transport oxygen around the bloodstream.

One of the reasons we need various metal ions in our diet, is to provide the ions needed to form biologically important coordination complexes. Compounds that will form coordination complexes with specific metal ions can be used to clear poisonous metals from the body, or to assist in the clean up of pollution due to metals. On some food labels, you can find EDTA (for ethylenediamine-tetraacetic acid). This compound forms coordination complexes with metal ions that, if left uncombined, can encourage spoilage of the food. It is also one of the medically useful ligands, used to treat mercury and lead poisoning, for example, and it is added to donated blood to bind to calcium ions and keep the blood stable.

Corey, Elias James

Professor Elias James Corey of Harvard University was awarded the 1990 Nobel Prize in Chemistry for his development of the theory and methodology of *organic synthesis*. Corey is one of the most influential contributors to the development of our ability to transform simple organic molecules into desired synthetic molecules. He therefore laid many of the foundations for the vast global efforts in synthetic chemistry, which every day create new drug candidates, new materials and new organic chemicals of every shape and size.

One of Corey's achievements was to develop the process of 'retrosynthetic synthesis'. This involves considering the structure of a 'target' chemical, which you want to synthesize, then working backwards to consider which bonds must be broken to simplify the target structure into precursors that can already be made by known techniques. Having undertaken that theoretical procedure, you can then begin to develop the chemical steps needed to build the target from the precursors you have identified. Corey also demonstrated that this logical way of analysing a problem in synthetic chemistry could be performed by computer. This has contributed significantly to the steadily increasing use of computers in developing new synthetic chemistry procedures.

Corey and his research team have achieved the synthesis of more than 100 naturally occurring products. This offers a means of making plentiful supplies of rare natural products, opens up chemical routes to modifying the products, and also helps us to understand how the products are made in nature. Corey's work has also developed some of the key synthetic procedures which thousands of other chemists have used to develop a wide range of medically and industrially useful products.

Corey's work continues at Harvard, where his research group currently focus much of their attention on the search for new synthetic catalysts for important organic transformations.

see also...
Organic Chemistry

Corrosion

orrosion can mean the degradation of any material by chemical reaction. Most commonly it refers to the slow disintegration of metals when they react with oxygen from the air, especially in the presence of water, perhaps accelerated by the presence of other 'corrosive' chemicals such as acids. The corrosion of metals is essentially a 'redox' reaction, involving the oxidation of metal atoms as they lose electrons to form metal ions. The electrons are transferred to whatever chemical the metal is reacting with.

The rusting of iron and steel (which is mainly iron) is the most familiar example of corrosion. Iron atoms react in the presence of oxygen dissolved in water to form Fe^{3+} ions bound within various forms of hydrated iron (III) oxide ($Fe_2O_3 \cdot nH_2O$). The iron oxide, which we see as 'rust' is more fragile than iron or steel, so it soon degrades into the flaking fissured structure of rusting metal. The corrosion of iron is a major problem, because it makes vulnerable parts of our vehicles, bridges, buildings, etc., begin to corrode as soon as they are built.

This is one of the disadvantages of living in an atmosphere rich in oxygen gas, which is chemically very reactive. We depend on that reactivity of oxygen, however, to react with fuel to get our vehicles moving, and to react with food to keep ourselves alive and moving. Oxygen reacting with metal, fuel and food are three separate aspects of oxygen's reactivity.

To slow or prevent corrosion, we can seal the metal off from contact with air and moisture using an unreactive covering such as plastic, or a less reactive metal such as tin. We can also discourage the loss of electrons from metal ions by making the metal negatively charged. This is called 'cathodic protection'. Another option is to attach any metal to be protected to a more reactive metal, which will corrode preferentially. Such 'sacrificial protection' is used to protect iron pipelines, using replenishable bins of magnesium scrap linked to the pipeline at intervals.

see also...

Oxidation, Reduction

Covalent Bonds

The almost infinite variety of chemical compounds is traditionally divided into two categories: those held together by covalent bonds and those held together by ionic bonds. Covalent bonds form when electrons become *shared* among atomic nuclei. The nuclei become bonded together into a molecule due to the forces of attraction between the nuclei and the shared bonding electrons. Hydrogen molecules (H_2) offer the simplest example, forming when the single electrons of two hydrogen atoms become shared between the nuclei of the two atoms. The hydrogen molecule is like an atom with two nuclei, and the electrons occupy a new 'molecular orbital' surrounding both nuclei rather than the individual atomic orbitals they came from. One pair of shared electrons corresponds to a single 'covalent bond', depicted by a single line in structural formulas. So H_2 is shown as H–H. Many molecules include double covalent bonds (two pairs of shared electrons), such as carbon dioxide (CO_2) shown as O=C=O. Others have triple covalent bonds (three pairs of shared electrons), such as N_2 molecules (N≡N).

When covalent bonds form between atoms with substantially different attractions for the bonding electrons (different electronegativities), the electrons become shared 'unequally'. In the O–H bond, for example, the electrons are drawn more strongly to the highly electronegative oxygen atom. This causes a 'polar covalent bond' to form, with the oxygen having a slight negative charge (δ-) due to its relative excess of electrons, and the hydrogen having a slight positive charge (δ+) due to its relative deficiency in electrons. Polar covalent bonds greatly influence molecules' reactivities.

Bonding theory is a complex and controversial area of chemistry. Seemingly very different models of bonding are used by chemists to describe the same basic phenomenon.

> ### see also...
> *Electron Configuration, Electronegativity, Electron Orbitals, Ionic Bonds, Lewis Structures, Molecular Bonding Theories, Molecules, Polar Covalency*

Dalton, John

John Dalton (1766–1844) was a chemist and physicist based in Manchester who is remembered as one of the founding fathers of our modern understanding of chemistry. He was interested in the composition of substances, and wondered if their major components were combined together in specific ways, or just loosely mixed. He investigated various substances in which two elements form more than one type of compound. For example, there are three compounds of nitrogen and oxygen. One has nitrogen and oxygen combined in a 1 to 1 ratio; in another the ratio is 1 to 2, while in the third it is 1 to 3. Results like these revealed a general rule: *when the same elements combine within more than one compound, they do so in fixed whole number proportions.* This is the law of multiple proportions.

Dalton speculated that these results could be explained by the existence, for each element, of some basic particle with a specific mass that would be the smallest part of every element. In 1803 he presented his ideas before the Manchester Literary and Philosophical Society. They are known as Dalton's atomic theory, and can be summarized as follows:

● Every substance is made of atoms.
● Atoms are indestructible.
● Atoms of any one element are identical.
● Atoms of different elements differ in their masses.
● Chemical changes involve rearranging the attachments between atoms.

Dalton's atomic theory forms a keystone in the foundations of chemistry; but if we look at it in the light of modern knowledge, we can see that not all of it is correct. Atoms are not indestructible or indivisible. Atoms of any one element are not all identical, but can exist in the form of different isotopes containing different numbers of neutrons. Dalton's atomic theory was nevertheless a great step forward in chemistry. It focused attention on the crucial idea that compounds are composed of discrete particles – atoms – combining in fixed proportions.

see also...

Atoms, Elements, Isotopes

Deoxyribonucleic Acid (DNA)

DNA is the chemical our genes are made of. It adopts the celebrated 'double helix' configuration which has become a universal symbol for biology and genetics. The analysis of DNA is the centrepiece of The Human Genome Project, aimed at fully deciphering the secrets of this giant chemical that controls the structure and activities of life.

Chemically, DNA is a 'nucleic acid', composed of a 'sugar-phosphate backbone' of alternating phosphate and deoxyribose sugar groups, with nitrogenous bases attached to this backbone in variable sequences. Two strands of DNA can wind around each other to form a double helix, provided they have appropriate 'complementary' sequences of bases. The four bases in DNA are adenine (A), guanine (G), thymine (T) and cytosine (C). An adenine on one strand of DNA can bind loosely to a thymine on another strand, forming the A–T base pair. Guanine and cytosine can form a G–C base pair. The base pairs are held together by attractions known as hydrogen bonds between slightly positive ($\delta+$) and slightly negative ($\delta-$) charges found on the bases.

The double-helical structure of DNA embodies two chemical properties fundamental to life: the ability to contain coded chemical information, and the ability to be copied into identical replicas. The sequence of bases in the DNA of a gene determines the sequence in which amino acid molecules will be linked together to form a protein 'encoded' by the gene. This base sequence is also called a nucleotide sequence, since each unit of base, sugar and phosphate is a nucleotide. By controlling which proteins are made in living things, genes control much of what happens in living things. The copying of DNA is achieved when the double helix unwinds, allowing enzymes to link together new complementary strands on the template of each original strand, in accordance with the rules of base-pairing discussed above.

see also...

Enzymes, Polar Covalency, Proteins, Ribonucleic Acid

Electrolysis

Electrolysis is the breaking down or 'decomposition' of a chemical using electricity. It is most commonly applied to molten ionic compounds and to solutions. It can be used to release a desired element from a compound, and is the basic technique of the electroplating industry.

The reactive metallic element sodium can be produced by the electrolysis of molten sodium chloride (NaCl). The positive sodium ions (Na^+) are drawn towards the negatively charged cathode, where they collect electrons to form sodium atoms (Na). The negative chloride ions (Cl^-) are drawn towards the positive anode, where they release electrons to form chlorine gas.

electrons. The oxidation occurs when negative ions lose electrons:

$$Na^+ + e^- \rightarrow Na \qquad \text{reduction}$$
$$2Cl^- \rightarrow Cl_2 + 2e^- \quad \text{oxidation}$$
$$\overline{2Na^+ + 2Cl^- \rightarrow 2Na + Cl_2 \quad \text{redox}}$$

Electrolysis is used for the industrial manufacture of several metals, including sodium, aluminium and copper; and gases such as hydrogen and chlorine. The metals always form at the cathode, because their ions are positively charged. For this reason, positively charged ions are called *cations*. Negatively charged ions, such as Cl^-, which move to the anode, are called *anions*.

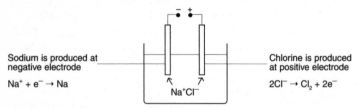

Sodium is produced at negative electrode

$$Na^+ + e^- \rightarrow Na$$

Na^+Cl^-

Chlorine is produced at positive electrode

$$2Cl^- \rightarrow Cl_2 + 2e^-$$

The electrolysis of molten sodium chloride

Electrolysis is a 'redox' process, involving reduction and oxidation half-reactions. The reduction step occurs when positive ions gain

> ### see also...
> *Oxidation, Reduction*

Electron Configuration

The building blocks of chemicals are atoms, which consist of a positively charged nucleus surrounded by moving electrons. The way in which the electrons are distributed among different energy levels is known as an atom's electron configuration (or 'electron arrangement' or 'electronic structure'). This is a major factor influencing the reactivity and behaviour of an atom, since electrons are the outermost parts of atoms that interact most strongly when atoms encounter other atoms, molecules or ions.

The electron configuration of an atom is presented as a series of numbers, indicating how many electrons are in each energy level. Sodium atoms, for example, have the electron configuration 2,8,1 – they have two electrons in the first energy level (or 'shell'), eight in the second and one in the third.

There are crucial links between an atom's electron configuration and its location in the Periodic Table. The period number, corresponds to the number of occupied energy levels, so we can tell that sodium must be found in Period 3. Also, atoms in the same main group of the Periodic Table have the same number of electrons in their outer energy level. This explains why elements in the same main group tend to share broadly similar chemical characteristics.

When atoms gain or lose electrons to form ions, their electron configuration changes into that of the ion. Sodium ions (Na^+) are formed when sodium atoms lose an electron, so the ions have the electron configuration 2,8. A chloride ion (Cl^-) has the electron configuration 2,8,8, since it is formed when a chlorine atom (2,8,7) gains one electron.

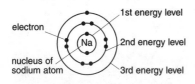

electron

nucleus of
sodium atom

1st energy level

2nd energy level

3rd energy level

A sodium atom electron configuration 2,8,1

see also...

Atoms, Electron Orbitals, Ionic Bonds, Periodic Table

Electronegativity

Chemistry can be viewed as a battle among atomic nuclei for the electrons that move around the nuclei. There are forces of attraction between the positively charged nucleus of an atom and the negatively charged electrons around it, and also around neighbouring atoms, molecules or ions. Molecules are formed when electrons become shared among different nuclei due to these forces of attraction. Ions are formed when electrons are pulled from one atom onto another, due to the greater electron-attracting power of one nucleus. Electrons repel one another, since they each have a negative charge, so the mutual repulsions among electrons counteract the attractive forces drawing them towards the 'competing' atomic nuclei. Thus, chemistry proceeds, with atoms, molecules and ions moving around and colliding in ways that cause the redistribution of electrons as they are pushed and pulled by the forces of attraction and repulsion acting on them.

The extent to which any type of atom can attract the shared electrons involved in covalent bonds is known as the atom's electronegativity, measured on a simple numerical scale. Although defined in terms of the attraction for electrons in covalent bonds, an atom's electronegativity gives a good impression of its ability to attract electrons in general. The most electronegative atom is *fluorine*, with an electronegativity of 4.0. The least electronegative is *francium*, with a value of 0.7. The electronegativity is influenced by two main factors: the size of positive charge on the atom's nucleus, and the 'screening' effect of the other electrons in an atom, which tends to shield the nuclear charge from having its maximal effect on bonding electrons. In general, electronegativity increases moving upwards and to the right of the Periodic Table.

Differences in electronegativity underlie the important phenomenon of 'polar covalency' in which covalent bonds become 'polarized' into δ^+ and δ^- poles.

see also...

Atoms, Covalent Bonds, Ionic Bonds, Polar Covalency

Electron Orbitals

The particles that chemicals are made of – atoms, molecules and ions – are composed of positively charged atomic nuclei surrounded by moving electrons. The electrons occupy specific regions of space known as electron orbitals. Each orbital can accommodate a maximum of two electrons. Electrons possess a property loosely described as 'spin', and more technically as *angular momentum*. A pair of electrons occupying the same orbital must have opposite electron spins.

In atoms, there are four kinds of 'atomic orbitals': *s* orbitals are spherical, *p* orbitals are dumbbell shaped, while *d* and *f* orbitals have more complex structures. The orbitals of any one kind in a particular energy level are smaller than the same kind in higher energy levels, but of the same general shape. The shapes of electron orbitals result from the wave-like nature of electrons. If we regard the electron as a tiny particle, however, we should imagine it moving around within its much larger orbital, like an aircraft moving within a flight path in the sky. That analogy also matches the fact that the orbitals themselves have no real physical existence – they are just the places where electrons can be found. The shapes of occupied orbitals can now be detected and indirectly visualized, however, and long-standing theoretical predictions of orbital structure have been experimentally confirmed in recent years. Hybrid orbitals can also form when several orbitals are combined. In carbon atoms, for example, the *s* orbital in the highest energy level often combines with three *p* orbitals to form four *sp^3* hybrid orbitals arranged tetrahedrally around the atom.

Electrons in molecules can occupy 'molecular orbitals' which are spread across the entire molecule according to the 'molecular orbital' model of bonding, or localized around neighbouring atoms, according to the 'valence bond' model.

see also...

Atoms, Electron Configuration, Molecular Bonding Theories, Quantum Chemistry, Sub-atomic Particles

Electrophiles and Nucleophiles

Chemical reactions involve electrons moving around and being redistributed, pushed and pulled by the attractions of the various atomic nuclei near to them and the repulsions of neighbouring electrons. When we examine the mechanisms of reactions we find that some chemical species clearly act as donors of electrons, while others accept electrons. Chemical groups that accept electrons are called electrophiles (electron-lovers) due to their attraction to the negative charge on electrons. Electrophiles are often positively charged species, in which case the reasons for their attraction for electrons are obvious, since positive charge attracts negative charge. Many other electrophiles, however, do not carry a full positive charge, but have

nucleophiles (nucleus-lovers) due to the attraction between the electrons they donate and the positive nuclei of the atoms that accept these electrons. Nucleophiles are often negatively charged species, which explains their ability to donate electrons (the carriers of negative charge). Many nucleophiles, however, do not carry a full negative charge, but have electron-rich regions, such as the slightly negative ($\delta-$) end of a polar covalent bond.

Study of the mechanism of chemical reactions, especially those involving organic compounds, often reveals electrons moving from nucleophiles to electrophiles in complex patterns. Chemists keep track of these electron movements using 'curly arrows'.

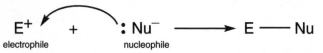

$$E^+ \quad + \quad :Nu^- \longrightarrow E{-\!-\!-}Nu$$

electrophile   nucleophile

The curly arrow represents an electron-pair moving to form the new bond. When a single electron moves, a 'fish-hook' arrow is used:

electron-deficient regions, such as the slightly positive ($\delta+$) end of a polar covalent bond. Chemical groups that donate electrons are called

see also...

Polar Covalency

Elements

Earth, water, wind and fire were the original 'elements', highlighted when people tried to identify the most fundamental aspects of Nature. We still talk about 'the elements' in that ancient way, when describing the weather and other forces Nature throws at us. In its modern context, however, the word elements refers to substances that contain only one type of atom. The elements are listed in the Periodic Table of the Elements, which grows familiar to most of us through our schooldays. There are 91 naturally occurring elements on Earth, including such familiar substances as oxygen, iron, aluminium, copper and tin. Scientists have artificially created atoms of around 20 other elements. These human-made elements are called the 'transuranium elements', because their atoms are larger than those of uranium. Many human-made elements are very unstable and short-lived, but plutonium, the raw material of the hydrogen bomb, is a notable exception. Miniscule amounts of naturally occurring plutonium may exist, but it is generally regarded as a human-made element.

Most elements are solids at room temperature and pressure. Eleven are gases: hydrogen, nitrogen, oxygen, fluorine, chlorine and the unreactive 'noble gases' (helium, neon, argon, krypton, xenon and radon). Two are liquids: mercury and bromine.

We use pure elements in many ways. For example, iron is the major component of steel, made by mixing small quantities of carbon and other elements into the iron. Copper forms the electrical wiring in our homes and workplaces. Silver and gold are prized elements used in jewellery.

The other major significance of the elements is that they are the raw materials that form the infinitely greater variety of chemical 'compounds'. Each compound contains two or more elements chemically bonded together. The 100 or so elements can combine to form an almost limitless number of chemical compounds.

see also...

Atoms, Compounds, Noble Gases, Periodic Table

Energy

Energy is defined as 'the capacity to do work', and 'work' is done whenever a physical force is applied through some distance. Every chemical reaction is accompanied by an energy change. Energy is released from 'exothermic' reactions, such as the burning of fuel. Energy is taken in by 'endothermic' reactions, such as photosynthesis. Energy commonly enters or leaves chemicals in the form of heat and/or light. The energy stored within chemicals is called 'chemical energy'. All types of energy, however, essentially involve just three fundamental forms of energy: kinetic energy, potential energy and electromagnetic radiation.

Kinetic energy is the energy objects possess due to their movement – as they speed up their kinetic energy increases, and as they slow down it decreases. *Potential* energy is stored energy embodied within an object's position. A rock held high above the earth, for example, has potential energy because the force of gravity would make the rock fall. An electron some distance from an atom's nucleus has potential energy because the force of attraction between the negative electron and positive nucleus would pull the electron towards the nucleus. *Electromagnetic radiation* is energy moving through space in the form of electromagnetic waves such as light.

Chemicals contain kinetic energy as a result of the movement of their particles (atoms, molecules, ions, nuclei and electrons), and potential energy as a result of the positions of their electrons and nuclei. During a chemical reaction, the nuclei and electrons of the chemicals shift into new arrangements, which embody a different amount of energy than they began with. That is the fundamental reason for the release or uptake of energy accompanying every chemical reaction. Light and other forms of electromagnetic energy can be absorbed by atoms and molecules in a way that moves electrons into higher energy orbitals. This is why chemicals have different colours, because they absorb different wavelengths of light.

> ### see also...
>
> *Electron Orbitals, Entropy, Photosynthesis*

Entropy

Why do things change? Why does chemistry happen? Why do plants and animals grow, then die and decay? These are big questions, but scientists can answer them all using the idea of entropy. Things change because the random movement of particles and energy inevitably causes energy and matter to *disperse*. This direction of natural change is defined in terms of a property of each system known as its entropy, and the entropy of the universe always increases, overall, during any natural change. That rule is called the Second Law of Thermodynamics. Entropy can be defined in strict mathematical terms, but we can understand it simply by appreciating that increasing entropy is associated with the automatic dispersal of energy and matter.

For example, why does a gas expand to fill the volume available to it? This happens because the random movement of the gas particles causes them to disperse into the available space. Having dispersed, the random movement of the particles also ensures they will never become concentrated in just one little part of

their container. Why do chemical reactions occur? They occur because, as they proceed, energy and matter automatically move towards more dispersed arrangements, overall, than they were in to begin with.

Although spontaneous chemical reactions are ones that are accompanied by an increase in entropy, naturally non-spontaneous reactions can still be made to happen. We can force otherwise non-spontaneous reactions to occur provided they are *coupled* to another spontaneous process, so that entropy still increases overall. To use a simple analogy, a rugby ball will never rise spontaneously up in the air; but it can be made to do so if it is coupled to the swinging of a heavy boot on the end of a rugby player's leg. In a similar way, the chemistry that lets plants grow would never happen on a cold dark Earth with no sun; but it is made to happen when the chemicals in plants have the energy of sunlight continually dispersing through them.

see also...

Energy, Photosynthesis

41

Enzymes

Almost every crucial chemical reaction within a living organism is catalysed by a specific enzyme. The chemical reactions of life would not proceed at the necessary rate without the assistance of enzymes, so enzymes are the crucial catalysts that make life possible. Enzymes are large protein molecules whose precise folded structure creates 'active sites' to which certain chemicals can bind. A particular reaction involving these bound chemicals is catalysed by the arrangement of chemical groups that form the enzyme's active site. The products of the reaction are then released, allowing the enzyme to go through the catalytic process many more times. The chemicals an enzyme binds to and acts on are known as the 'substrates' of the enzyme-catalysed reaction. The equation below summarises the three-step process of enzyme (E) binding to substrate (S) to form an enzyme-substrate complex (ES), which changes into an enzyme-product complex (EP), which then releases the products (P) and returns the enzyme to its original state.

$$E + S \rightleftharpoons ES \rightleftharpoons EP \rightleftharpoons E + P$$

For example, digestive enzymes bind to carbohydrates, fats and proteins in our diet and combine them with water molecules to break them into smaller nutrients able to be absorbed into the blood. Some enzymes are assisted in their tasks by small molecules called 'co-enzymes' which must bind to the enzyme in order to form the active site. Many vitamins are essential nutrients because they either act as co-enzymes directly, or are converted into co-enzymes once they enter the body. Other enzymes, called 'metallo-enzymes' require specific metal ions to form part of their active site. This explains many of our 'trace mineral' dietary needs, because we need ions such as copper, selenium or zinc to form metallo-enzymes.

Some enzymes are purified from living things and used as industrial or commercial reagents. Biological washing powder, for example, contains enzymes able to digest difficult biological stains made by such things as food, blood or grass.

see also...

Catalyst, Minerals, Proteins

Equations

A chemical equation summarises what happens during a chemical reaction using chemical formulas and numbers. The equation conveys information that could readily be said in words, but it conveys the information in a more concise form.

For example, in a blast furnace molten iron oxide reacts with carbon monoxide gas to form molten iron and carbon dioxide gas. The formula for the iron oxide is Fe_2O_3, the formula for iron is Fe, the formula for carbon monoxide is CO, and the formula for carbon dioxide is CO_2. The iron oxide and carbon monoxide react in a 1:3 ratio, producing iron and carbon dioxide in a 2:3 ratio. All of that information can be conveyed in just one line, using the following equation:

Fe_2O_3 (l) + 3CO (g) $\rightarrow$ 2Fe (l) + 3CO_2 (g)

After each formula, (l) indicates a liquid, while (g) indicates a gas. We can also use (s) for a solid and (aq) for an 'aqueous' (water-based) solution.

Every equation must be 'balanced', with all of the atoms present at the start of the reaction also present at the end. This is because atoms are not created or destroyed during a chemical reaction, they are simply rearranged.

For example, when methane (CH_4) burns it combines with oxygen (O_2) in the air to produce carbon dioxide (CO_2) and water (H_2O). We can set out the formulas as follows:

CH_4 (g) + O_2 (g) $\rightarrow$ CO_2 (g) + H_2O (l)

That equation is unbalanced, because there are four hydrogen atoms in the reactants on the left of the equation, but only two hydrogen atoms in the products, on the right. Also, there are two oxygen atoms in the reactants but three in the products. We can balance the equation using numbers to indicate that two oxygen molecules react with each methane molecule, to produce one carbon dioxide molecule and two water molecules:

CH_4 (g) + 2O_2 (g) $\rightarrow$ CO_2 (g) + 2H_2O (l)

In balancing an equation, we discover the ratio in which chemicals react and are produced in Nature.

see also...

Formula

Equilibrium

The red colour of meat is largely due to the presence of iron-containing protein molecules called myoglobin, packed inside muscle cells. The myoglobin molecules serve as an oxygen store, holding onto oxygen when it is in good supply then releasing it when the cells need it quickly, such as when we run. Myoglobin is able to perform this useful function due to an important phenomenon called chemical equilibrium.

The existence of equilibrium relies on the fact that many chemical processes are readily reversible. When oxygen is mixed with myoglobin, for example, some of the oxygen molecules will bind to myoglobin; but then some of the bound oxygen molecules will begin to be released. The binding reaction proceeds reversibly, in both directions at the same time, depicted using a double-headed reaction arrow:

Under any particular conditions of oxygen concentration, myoglobin concentration, temperature and pressure, the system will soon settle into a state in which *both forward and reverse reactions are proceeding at the same rate*. This is known as the state of equilibrium. At equilibrium the concentrations of the chemicals involved are not changing, but there is a *dynamic* equilibrium, because the reaction is proceeding in both directions at equal rates that cancel out overall.

Reversible reactions shifting between different positions of equilibrium in response to changes in chemical concentrations and conditions underlie many of the exquisite control systems of the chemistry of life. Many industrial reactions also involve equilibrium processes.

$$O_2 + Mb \rightleftharpoons MbO_2$$
$$\text{Oxygen} + \text{Myoglobin} \rightleftharpoons \text{Myoglobin-Oxygen}$$

see also...

Chemical Reaction

Esters

Many of the natural scents and flavourings in foods are compounds called esters, as are many of the synthetic perfumes manufactured for our pleasure and personal hygiene. The polyester fabrics used in everyday

shown below, is formed in a condensation reaction between a carboxylic acid and an alcohol. Polyesters, have many ester linkages holding various 'monomer' units into a long polymer chain.

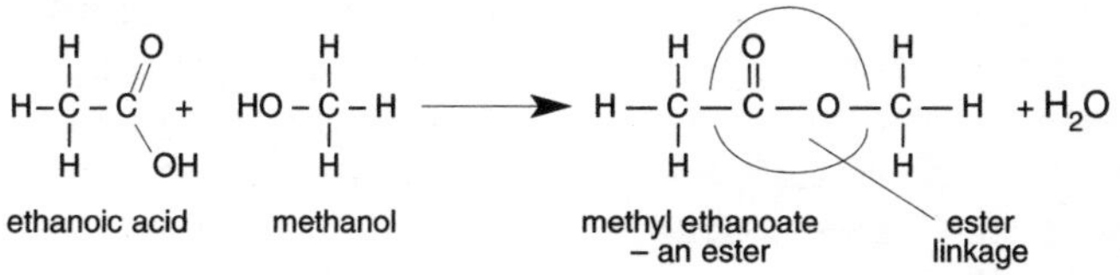

ethanoic acid   methanol   methyl ethanoate   ester
– an ester   linkage

clothing are giant esters. Esters play many important roles in living things and are one of the most versatile categories of chemical available for synthetic chemists to use when creating new materials, medicines or other synthetic substances. Esters come in a huge variety of chemical structures. The common feature they

Fats and oils, such as animal lard and cooking oil, are 'triesters' made when three hydroxyl (OH) groups in glycerol combine with the carboxyl (COOH) groups on three long-chain 'fatty acid' molecules. Soaps are made when these ester linkages are broken (hydrolysed) to form salts of the fatty acids that are released.

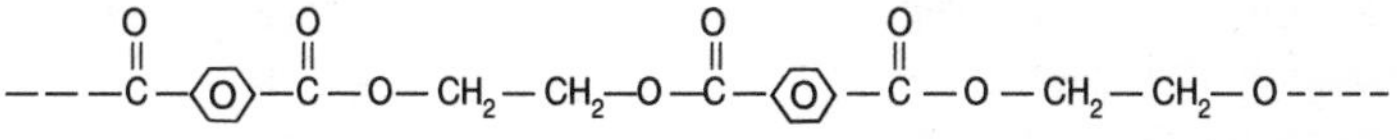

The repeating structure of a polyester.

all share is that they possess at least one 'ester linkage'; a simple group of one carbon and two oxygen atoms bonded on either side to two further carbon atoms to hold the ester molecule together. An ester linkage,

see also...

Condensation and Hydrolysis Reactions, Fats and Oils, Functional Groups, Polymers, Soaps and Detergents

Explosives

Explosive chemicals are used widely in our society, most obviously in weapons, but also for peaceful purposes such as quarrying, road-building, demolition, rocket propulsion and in fireworks. Explosives are chemicals in a 'metastable' state. This means they are stable enough not to react together spontaneously, but sufficiently unstable to be jolted into a violent self-sustaining reaction if provided with a little energy to make the process start. The ensuing reaction must release energy very fast, to cause the expansion of hot gases and reacting debris that is what we really mean by an explosion. The most rapid and uncontrolled explosions are the 'detonations' associated with weapons, quarrying, demolition and fireworks. The chemicals used for these fast explosions are called 'high explosives'. More controlled explosive reactions can be used to blast spacecraft into orbit. In these cases the chemicals are described as 'low explosives'.

The original explosive was gunpowder, a mixture of potassium nitrate, sulphur and charcoal (essentially carbon). Then nitrocellulose (cotton powder) was made by treating cotton (a rich source of cellulose) with nitric acid and sulphuric acid. Next came nitroglycerine, made by treating glycerol (released from fats and oils) with nitric acid and sulphuric acid. Nitroglycerine is dangerously unstable, but is converted into the more stable form, dynamite, by absorbing it into the inert solid silica (a form of silicon dioxide). Nowadays a great variety of explosives is available for specific uses, including many 'plastic explosives' that can be moulded into any shape. The best known modern explosives are probably trinitrotoluene (TNT), shown below and Semtex.

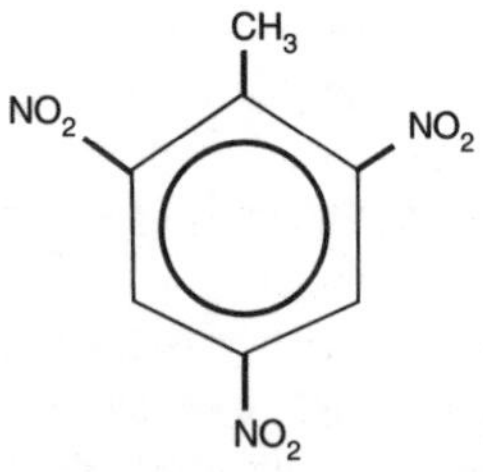

Trinitrotoluene (TNT)

Fats and Oils

The fats and oils familiar to us as food components are composed of 'triglyceride' molecules. These form when the hydroxyl (OH) groups of glycerol form ester linkages with the carboxyl (COOH) groups of three long-chain fatty acids:

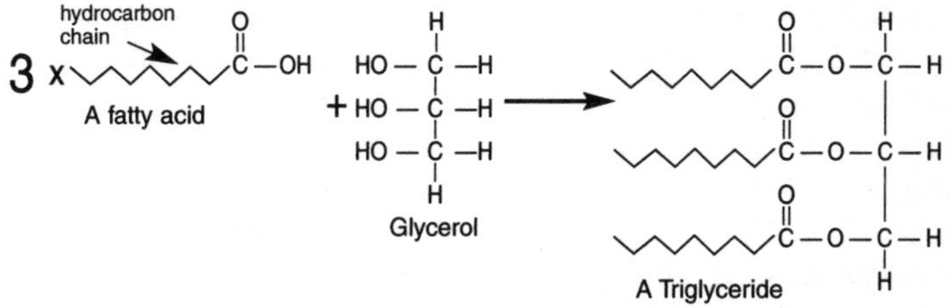

The fatty acids can be identical or different, depending on the triglycerides concerned. Fats are solids at room temperature whereas oils are liquids. So fats have higher melting points than oils. The main chemical cause of the differing melting points is the presence of more carbon-carbon double bonds (C=C) within the fatty acids of oils. These bonds introduce kinks in the fatty acid hydrocarbon chains that make it more difficult for the triglycerides to settle into the ordered structure of a solid. Fats containing C=C double bonds are 'unsaturated fats', while those with no C=C bonds

are 'saturated'. Monounsaturated fats contain no more than one C=C bond in each fatty acid, while polyunsaturated fats contain more than one. These terms are familiar to us from their use on food labels. These labels may also mention '*trans*' fatty acids, which have hydrogen atoms on opposite sides of the double bonds, as opposed to '*cis*' fatty acids with hydrogens on the same side:

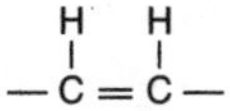

cis-configuration

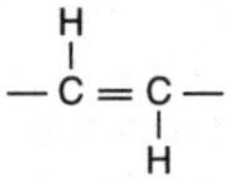

trans-configuration

see also...

Esters, Functional Groups, Hydrocarbons

Fertilizers

Plants require appropriate sources of various elements to promote healthy growth. Some of these elements are available in abundance, such as the carbon which plants get from carbon dioxide (CO_2) in the air, and hydrogen and oxygen which they get from water (H_2O) drawn up through their roots. Supplies of other elements can be quite limited, depending on the soil the plants are growing in. For example, the restricted availability of nitrogen, phosphorous and potassium in forms that can be used by plants, can limit the growth of many crops. Fertilizers are chemicals that are applied to soil in order to supply essential elements and therefore increase plant growth and crop yields. Six commonly used fertilizers are ammonia – NH_3, urea – $(NH_2)_2CO$, ammonium nitrate – NH_4NO_3, ammonium phosphate – $(NH_4)_3PO_4$, ammonium sulphate $(NH_4)_2SO_4$ and potassium nitrate – KNO_3 . These are all 'inorganic fertilizers', in the sense that they are inorganic compounds. They are compounds that are soluble in water, since they must dissolve in order to be effectively taken up from moist soil by the roots of plants.

Other, more natural, fertilizers are supplied by the rich mixture of compounds within animal wastes ('manure'), rotting plant material ('compost') and dead or decomposing organisms of any form. These are known as 'organic fertilizers' because they are derived from carbon-based living organisms. These are the only fertilisers allowed in the production of so called 'organic' foods.

The manufacture of fertilizers is one of the main activities of the chemical industry. Their use is linked to a variety of environmental problems. If used in large quantities they can be washed into rivers and lakes to cause a problem known as 'eutrophication'. This is a complex ecological imbalance caused by the uncontrolled growth of algae, promoted by the fertilizers, leading to the death of water plants, the loss of dissolved oxygen due to the action of bacteria feeding on decaying plants, and the death of fish and other forms of life that depend on dissolved oxygen.

see also...

Nitrogen Cycle, Organic Chemistry

Formula

Each chemical compound can be represented by a formula, indicating the elements present in the compound and the proportions in which they are present. The elements are represented by their symbols, shown in the Periodic Table of the Elements. Almost everybody is familiar with the formula for water – H_2O – which indicates that water contains two hydrogen atoms for every oxygen atom. Note that subscript numbers in formulas, such as the 2 in H_2O, refer to the element preceding the number. We can also make these numbers refer to more than one element using brackets. For example, the formula for calcium hydroxide, a form of lime, is $Ca(OH)_2$. The 2 refers to both the oxygen (O) and hydrogen (H). So calcium hydroxide contains two oxygen atoms and two hydrogen atoms for every calcium atom.

In calcium hydroxide, the oxygen and hydrogen atoms are bonded together to form a hydroxide ion (OH^-), which explains why we use the formula $Ca(OH)_2$ rather than CaO_2H_2. The latter version is still a valid formula. It is called the 'empirical' formula, which simply indicates the elements present and the proportions of each without embodying any further information about the compound. Compounds that exist in the form of molecules can have a 'molecular formula', indicating exactly which atoms are present in each molecule, which looks very different from the empirical formula. For example, octane, a hydrocarbon found in oil, consists of molecules containing eight carbon atoms and 18 hydrogen atoms. Its molecular formula is therefore C_8H_{18}. Its empirical formula, however, is C_4H_9, since that is the simplest whole number ratio that can be used to indicate the proportions in which the atoms are present.

Some compounds contain water molecules incorporated into their structure as 'water of crystallization'. This is indicated using a dot, followed by the formula for water. For example, $CuSO_4 \bullet 5H_2O$ is 'hydrated copper sulphate', with five water molecules for each $CuSO_4$ formula unit.

see also...

Atoms, Equations, Molecules, Periodic Table

Free Radicals

An atom or molecule that carries at least one unpaired electron in its outer energy level is called a free radical. Many free radicals are extremely reactive substances, because a single unpaired electron tends to pair up with another electron pulled from another chemical species. Free radicals are often symbolised with a dot indicating the unpaired electron, such as O•, representing the oxygen free radical (just a free oxygen atom). Free radicals can be grouped into three classes: *atoms* (such as O•, H•, Cl• and F•), *organic free radicals* (such as •CH_3 and •C_5H_6) and *inorganic free radicals* (such as •OH, •CN and ClO•). Not all are highly reactive. For example, nitrogen oxide, NO•, is relatively stable.

The reactivity of many free radicals makes them a danger to living things. Free radicals are believed to be involved in the onset of some cancers and in the slow accumulation of chemical damage associated with ageing. This is because their reactivity allows them to react with and therefore damage other chemicals of the body, including such vital ones as the DNA of our genes. The body has complex defensive systems to protect against free radicals which form spontaneously from water, oxygen and chemicals in our food. Free radicals are also involved in the reactions associated with pollution problems, such as the formation of photochemical smog. The formation of many free radicals is initiated by the absorption of light or electromagnetic radiation of other wavelengths. The absorbed energy can split the electron pair of a covalent bond, for example when a chlorine molecule (Cl_2) is broken to yield two Cl• free radicals. The chemical industry exploits the reactivity of free radicals to initiate many important industrial reactions. Some drugs are designed to generate free radicals that will kill diseased tissue, such as cancer.

One of the reasons we are encouraged to eat plenty of fresh fruit and vegetables is because these foods contain natural chemicals which can combine with and therefore 'mop up' free radicals produced in the body.

see also...

Atoms, Electron Configuration, Smog

Fuels

Any chemical reaction that releases energy could, in principle, be used to provide heat, light or make machinery and vehicles operate. These are the things we ask of the chemicals we call fuels. The original fuel was dead plant material, especially wood. Then the industrial revolution and the development of modern society was made possible by the exploitation of coal, oil and natural gas. These are called 'fossil fuels' because they are derived from ancient plant and animal life. Our dependence on fossil fuels is linked with various problems. First, they are 'non-renewable'. When they are gone we will have to find alternatives. Second, the fossil fuels release carbon dioxide as they burn. The increase in carbon dioxide in the atmosphere is one of the main contributors to the global greenhouse effect, suspected of causing undesirable global warming. These problems have led to increasing efforts to find alternative fuels.

Some of the most practical alternatives involve a return to exploiting plant material. Rather than simply burning the material, however, it may be converted into convenient alternatives to oil such as *ethanol* (C_2H_5OH). Sugar cane is fermented to produce ethanol to power vehicles in Brazil. The ethanol is a renewable fuel since we can grow sugar cane to generate ethanol as fast as the ethanol is consumed. Although burning the ethanol does release carbon dioxide, an equivalent amount of carbon dioxide is taken up by the sugar cane as it grows. So ethanol is neutral in its effect on global warming, unlike the fossil fuels.

Hydrogen (H_2) might be the perfect fuel because when it burns to release energy water is the only product, so there are no pollution problems. The hydrogen can be released from water, powered by sunlight. If that process could be made efficient and industrially feasible we could generate plentiful pollution-free fuel by endlessly splitting water then reforming it. Burning hydrogen to produce electricity might be more practical than using it directly as a fuel.

see also...

Global Greenhouse Effect,
Photovoltaics

Fullerenes

For many years chemists believed there were only two distinct forms or 'allotropes' of the element carbon. These were *diamond*, in which each carbon atom is covalently bonded to four neighbouring atoms; and *graphite*, in which each atom is bonded to three neighbours. In 1985, however, an intriguing new way for carbon atoms to bond together was discovered. Allowing an electric discharge to arc between two graphite rods, surrounded by helium gas at high pressure, created molecules in which 60 carbon atoms were bonded into a soccer ball structure. The C_{60} molecule was called Buckminsterfullerene, due to its similarity to the geodesic domes designed by the architect Buckminster Fuller. The co-discoverers of C_{60} were rewarded with the 1996 Nobel Prize in chemistry. Buckminsterfullerene has since been joined by an increasingly varied group of related structures, known collectively as *fullerenes*. These include cage-like structures that can be larger or smaller than the C_{60} cage; and also tubes of bonded atoms, known as 'nanotubes' that can occur alone, or in concentric nested patterns.

Fullerenes have since been found to be abundant in nature, both on Earth and in interstellar space. Many chemists are trying to develop medical and technological applications for fullerenes and the related nanotubes. Some are trying to entrap drugs within the cage-like fullerenes, and use them as tiny delivery capsules for the medicines. Others are hoping to fabricate nanotubes into the wires of unbelievably small electrical circuits.

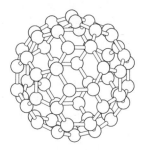

see also...

Molecular Electronics

Functional Groups

Functional groups are groups of atoms that display characteristic chemical properties, no matter which chemical structure they are attached to. The idea of functional groups is particularly useful in organic chemistry, because most organic chemicals can be regarded as composed of a simple hydrocarbon framework, with various functional groups incorporated.

Each functional group is found in a particular class of organic chemical. The alcohols, for example, contain the hydroxyl functional group. The

Some functional groups combine to form other functional groups. For example, the hydroxyl (OH) group reacts with the carboxyl (COOH) group to form the ester group (see diagram). In this process, the molecules carrying the hydroxyl and carboxyl groups become linked together by an 'ester linkage'. This allows the hydroxyl and carboxyl groups to be used to link together a variety of monomers to form polymers such as polyesters. Chemists use the reactivity of functional groups to manufacture new drugs and materials.

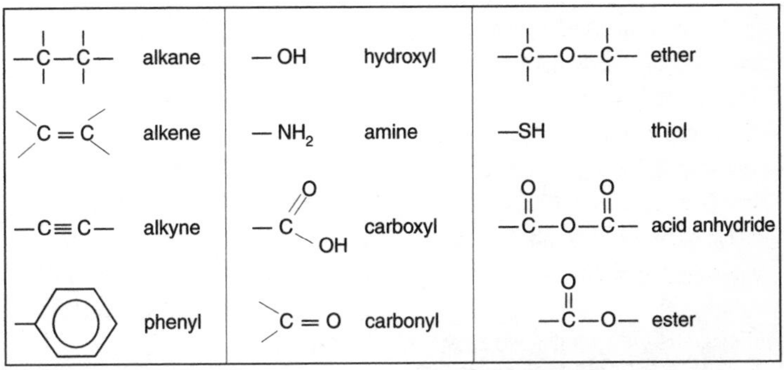

—C—C—	alkane	— OH	hydroxyl	—C—O—C—	ether
C = C	alkene	— NH$_2$	amine	—SH	thiol
—C≡C—	alkyne	—C(=O)OH	carboxyl	—C—O—C— (with two =O)	acid anhydride
phenyl ring	phenyl	C = O	carbonyl	—C(=O)—O—	ester

carboxylic acids contain the carboxyl functional group, and so on. Some important functional groups are shown in the diagram.

see also...

Esters, Hydrocarbons, Organic Chemistry, Polymers

Global Greenhouse Effect

The earth's atmosphere acts as an insulating blanket, keeping the earth warmer than it would otherwise be. This is due to various gases in the atmosphere which are transparent to the sunlight arriving from space, but can absorb infra-red heat energy reflected from the surface. In this regard, they act a bit like the glass of a greenhouse, which lets sunlight in but stops heat from getting out. The greenhouse analogy is not perfect, but it has led to the relevant gases in the atmosphere being called greenhouse gases, and to their overall effect being called the global greenhouse effect. This is an entirely natural effect that is important in keeping the planet a comfortable place for life. In recent years, however, the global greenhouse effect has become a major environmental issue due to the belief that we are enhancing it by releasing the pollution blamed for 'global warming'.

The main natural greenhouse gases are carbon dioxide (CO_2), methane (CH_4), nitrous oxide (N_2O) and ozone (O_3). Our burning of fossil fuels (coal, oil and natural gas), is releasing into the atmosphere huge quantities of carbon dioxide. This fast release of carbon dioxide is one of the major human interventions blamed for global warming. Human activities have also increased the levels of methane in the atmosphere, released from waste landfill sites and as a result of modern agricultural methods. The chlorofluorocarbon (CFC) molecules responsible for the depletion of the ozone layer are also many thousands of times more effective as greenhouse gases than carbon dioxide. So, although released in relatively tiny quantities they can have a disproportionately large effect.

Global warming due to the global greenhouse effect is a controversial topic because of the political and economic issues involved and also because there are some genuine scientific uncertainties. Remedies include reducing our reliance on fossil fuels and chemically trapping gases that are currently released unchecked.

see also...

Fuels, Ozone

Green Chemistry

The modern chemical industry has undoubtedly brought us great benefits, in the form of new materials and drugs for example, but it has environmental drawbacks as well. Our practice of chemistry is often cited as the villain in environmental debates. Pollution is largely caused by chemicals in the wrong places, or at inappropriate concentrations. Chemical wastes flowing from an industrial plant, choking photochemical smog in urban areas, the depletion of the ozone layer and the threat of global warming are all problems created by chemistry. The reputation of 'chemicals' has become so bad that some people crave chemical-free food and a chemical-free environment. By 'chemical-free', what they really mean is free of synthetic chemicals and undesirable accumulations of natural chemicals. Since all natural products and materials are composed of chemicals, the idea of chemical-free anything is an absurdity. Rather than turning our backs on chemical knowledge in the face of environmental problems, many chemists point out that chemical knowledge is our best weapon in the battle to solve the problems. Chemistry is needed to devise new and more environmentally friendly manufacturing and agricultural processes, and to devise more benign means of capturing and using energy. These endeavours have been given the general label of Green Chemistry.

The US Environmental Protection Agency defines the mission of Green Chemistry as:

"To promote innovative chemical technologies that reduce or eliminate the use or generation of hazardous substances in the design, manufacture and use of chemical products."

In other words, it aims to use chemistry in an environmentally friendly manner. Key principles of green chemistry are: prevent pollution rather than having to clean it up; develop technologies that minimise the use of hazardous compounds; minimise energy consumption in chemical processing; ensure products will break down naturally ('biodegrade') once they have been used.

> ### see also...
> *Global Greenhouse Effect, Ozone, Smog*

55

Halogens

The elements in Group 7A of the Periodic Table – fluorine (F), chlorine (Cl), bromine (Br), iodine (I) and astatine (At) – are known as the halogens. The term means 'salt makers', because these reactive non-metal elements will react with metals to form the range of compounds called salts. Everyday table salt, for example, is sodium chloride (NaCl). Other examples of halogen-containing salts are potassium chloride (KCl), sodium fluoride (NaF), lithium bromide (LiBr), and so on.

The halogen elements typically exist as diatomic molecules such as F_2, Cl_2 and Br_2. Astatine is generally regarded as not occurring naturally, the samples used for analysis having been human-made. The other four halogens are important to us both in their pure forms and when bonded within compounds. They occur naturally within compounds such as the various salts because the elements themselves are quite reactive, each atom readily accepting an electron to become a negatively charged ion.

Fluorine is a pale yellow gas at everyday temperatures and pressures. It is the most electronegative element known, and hence is extremely reactive. It is one of the few chemicals that can react with the famously unreactive noble gases. In the form of the fluoride ion (F^-), it is familiar to us as an additive in toothpastes, mouth rinses and some water supplies. Fluoride becomes incorporated into teeth and makes them more resistant to decay. *Chlorine*, a pale green gas, is also highly reactive. Its most familiar application is as a disinfectant used in water purification systems to sterilize both drinking water and swimming pool water. Chlorine compounds are also widely used in bleaches. *Bromine* is the only non-metal element that is a liquid at everyday temperatures and pressures, and one of only two elements that are liquid in these conditions (the other is mercury). *Iodine* is a dark brown solid, best known for the use of iodine solutions, and solutions of iodine compounds, as antiseptics.

see also...

Addition Reactions,
Electronegativity, Noble Gases,
Periodic Table

Heavy Metals

Anyone interested in environmental issues soon comes across the term heavy metals, usually in the context of pollution caused by these metals. Heavy metal is a label used rather loosely to describe metallic elements of 'relatively high atomic mass'. In practice, this usually means metallic elements from chromium upwards, in terms of mass. Some of the most significant heavy metals are chromium, manganese, iron, nickel, copper, cadmium, mercury and lead. The well known poisonous element arsenic, although technically a metalloid, is also generally regarded as a heavy metal. Some of the heavy metals are actually essential for life but many of them are extremely toxic. Some of those essential for life in small concentrations can be extremely toxic in high concentrations. Heavy metals in water and soil are a major focus of environmental concern. They can be particularly persistent environmental contaminants since, being elements, they cannot be degraded into simpler less toxic components. This is in contrast to toxic compounds, which are often readily degraded into less harmful compounds or their component elements.

The toxicity of heavy metals is generally due to the ability of ions of the metals to bind to important organic chemicals in living systems. The metal ions most commonly bind to oxygen, sulphur or nitrogen atoms in organic compounds and so prevent the compounds from playing their normal biological role. Important metal-containing enzyme molecules, for example, can be inactivated when heavy metal ions bind to them and so displace the natural metal ions from the enzyme's active site. Treatments for acute heavy metal poisoning generally involve administering drugs that will themselves bind to the heavy metal ions and carry them out of the body.

Heavy metal pollution has been a problem at least since the time of the Roman Empire, when lead plumbing resulted in lead pollution of water supplies.

see also...

Acid Rain, Coordination Complexes, Elements, Enzymes, Metals, Periodic Table

Hydrocarbons

Compounds composed solely of carbon and hydrogen are known as hydrocarbons. The simplest hydrocarbon is methane (CH_4), the main constituent of natural gas. Methane is found trapped above reservoirs of crude oil, which is a complex mixture of hydrocarbons based on straight chains, branched chains and rings of carbon atoms with hydrogen atoms attached. These include propane (C_3H_8) and butane (C_4H_{10}), which can be purified and used as fuels. Petrol and diesel fuels are mixtures of hydrocarbons mostly containing from six to twelve carbon atoms per molecule. Larger hydrocarbon molecules are found in lubricating oils. Asphalt and tar are composed of hydrocarbons with 20 or more carbon atoms per molecule. The larger hydrocarbons have higher boiling points than smaller ones. This allows crude oil to be separated into 'fractions', composed of mixtures of hydrocarbons with similar molecular sizes, in a fractional distillation column. The oil is heated to boiling and the fractions collected at different heights up the column, where they condense back into liquids as the temperature gradually falls.

Hydrocarbons are versatile raw materials used by the chemical industry to manufacture thousands of organic compounds. Many natural products contain hydrocarbon groups. Fats and oils, for example, include long hydrocarbon chains, as do all soaps and detergents. Hydrocarbons with carbon to carbon double bonds (C=C) are called 'unsaturated hydrocarbons'.

Those with no double bonds, and therefore with more hydrogen atoms attached, are 'saturated hydrocarbons' (since they are 'saturated' with hydrogen).

Octane – a saturated hydrocarbon

Butene – an unsaturated hydrocarbon

see also...

Aromatic Compounds, Fats and Oils, Fuels, Petrochemistry, Soap and Detergents

Intermolecular Forces

The forces of attraction between neighbouring molecules are generally much weaker than the covalent bonds that hold molecules together. These intermolecular forces, however, crucially influence a chemical's behaviour, including its melting point, boiling point and more subtle properties, many crucial for life. The most universal intermolecular forces are the 'van der Waals forces'. These are fleeting attractions between regions of slight negative charge on one molecule and regions of slight positive charge on another molecule, caused by the random movement of the electrons of the molecules. These weak forces of attraction occur between all molecules and all atoms. Their strength increases as the sizes of molecules or atoms increase.

Other intermolecular forces of attraction are set up between permanent slight negative ($\delta-$) regions within molecules and neighbouring slight positive ($\delta+$) regions of other molecules. When the $\delta+$ region is on a hydrogen atom, these forces are called 'hydrogen bonds'. They are the forces of attraction that hold the DNA double helix together, so they are crucial for the whole process of genetics and reproduction. Hydrogen bonds also attract water molecules together and keep most of the water on Earth in the liquid state rather than in the gas state. Without these intermolecular forces the world would be very different and life as we know it would be impossible.

The activities of proteins, another category of chemicals essential for life, also depend on a wide variety of intermolecular forces. These are the forces that allow enzymes to bind to the chemicals they act upon, and allow hormones and other biological signalling molecules to bind to and affect their receptor molecules on and inside individual cells. In industrial chemistry, intermolecular and interatomic forces are also vital, allowing gases to bind temporarily to the surface of catalysts, for example.

see also...

Deoxyribonucleic Acid (DNA),
Polar Covalency, Polar Molecules,
Proteins

Ionic Bonds

The almost infinite variety of chemical compounds is traditionally divided into two broad categories: those held together by covalent bonds and those held together by ionic bonds. Ionic bonds are formed when electrons are *transferred* from one atom to another, generating oppositely charged ions which are attracted to one another. Each atom that forms an ion tends, in the process, to acquire the stable outer electron arrangement of the nearest noble gas.

Sodium chloride (NaCl – common table salt) provides a simple example. A sodium atom can form a sodium ion (Na^+) when its sole outer electron is transferred onto a chlorine atom, which becomes a chloride ion (Cl^-). The Na^+ ion is left with the stable electron configuration of neon (2,8) while the Cl^- ion attains the stable electron configuration of argon (2,8,8). Ions with double or triple charges are also common. A calcium atom, for example, loses its two outer electrons to form a double positive Ca^{2+} ion. An oxygen atom, will gain two electrons to form a double negative O^{2-} ion. When ionic bonds form, atoms on the left of the Periodic Table generally lose electrons and form positive ions, while those on the right generally gain electrons and form negative ions.

The ionic bonds in solid ionic compounds hold the ions together in the rigid structure of a crystal lattice. Each compound's crystals have a distinctive shape, such as the cubic structure of sodium chloride crystals. Compounds held together by ionic bonds will conduct electricity when molten, or when dissolved in water, because their ions become mobile in these states, and able to serve as the carriers of electric current.

Classifying bonds within compounds as either ionic or covalent is a useful simplification of a more complicated reality. Many compounds are best described as having a certain percentage of 'ionic character' complemented by a percentage of 'covalent character'.

see also...

Covalent Bonds, Electron Configuration, Noble Gases, Periodic Table

Isomers

somers are compounds that
contain the same atoms arranged
in different ways. Three main

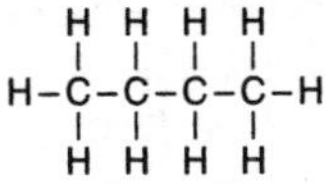

$$H-\overset{\overset{\displaystyle H}{|}}{C}-\overset{\overset{\displaystyle H}{|}}{C}-\overset{\overset{\displaystyle H}{|}}{C}-\overset{\overset{\displaystyle H}{|}}{C}-H$$
$$\underset{H}{|}\;\underset{H}{|}\;\underset{H}{|}\;\underset{H}{|}$$

butane (C_4H_{10})

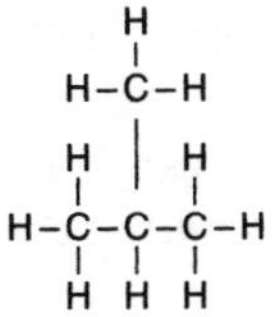

methyl propane (C_4H_{10})

Two structural isomers

types of isomers are structural
isomers, geometric isomers and
optical isomers. *Structural* isomers
have the same atoms bonded in a
different order.

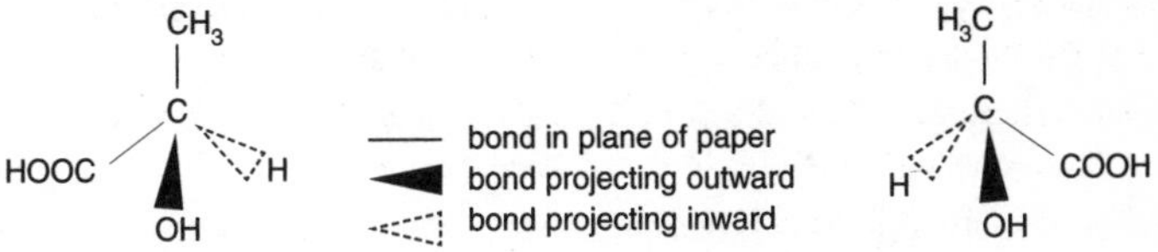

— bond in plane of paper
◄ bond projecting outward
◄--- bond projecting inward

Two optical isomers

Geometric isomers differ in the way in
which atoms or groups are arranged
on either side of a double bond. In the
example shown below, when the two

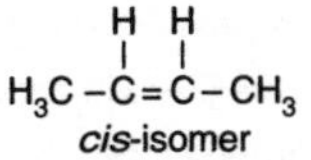

$$H_3C-\overset{\overset{\displaystyle H}{|}}{C}=\overset{\overset{\displaystyle H}{|}}{C}-CH_3$$

cis-isomer

$$H_3C-\overset{\overset{\displaystyle H}{|}}{C}=C-CH_3$$
$$\underset{H}{|}$$

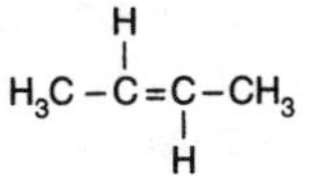

trans-isomer

Two geometric isomers

hydrogen atoms are on the same side
of the C=C double bond the *cis*-isomer
is formed. In the *trans*-isomer, the
hydrogen atoms are on opposite sides
of the double bond. This distinction is
possible because the arrangement of
electrons in double bonds prevents
rotation around the bond.

Optical isomers are mirror image
molecules; a subtle three-dimensional
difference that makes them non-
superimposable, like pairs of hands.

see also...

Compunds, Formula

61

Isotopes

Atoms of the same element which differ in the number of neutrons they contain are called isotopes. For example, all carbon atoms have 6 protons and 6 electrons, but different isotopes of carbon exist containing 6, 7 and 8 neutrons respectively. We can represent these using nuclide notation, with a superscript indicating the atom's mass number (number of protons plus neutrons) and a subscript indicating its atomic number (number of protons).

$$^{12}_{6}\text{C} \qquad ^{13}_{6}\text{C} \qquad ^{14}_{6}\text{C}$$

The three isotopes of carbon

Isotopes of any element behave in essentially the same way chemically, but they have different masses (of 12, 13 and 14 atomic mass units in the examples shown above) and can differ in their stability. Carbon-12, for example, is stable, while carbon-14 is a radioactive isotope that decays by releasing a beta particle (a fast-moving electron) when one of its neutrons changes into a proton, generating an atom of nitrogen-14.

Rare isotopes of elements can be used as 'labels' or 'tracers' in medicine and biochemistry, to follow the fate of particular chemicals or specific atoms within the human body or any organism. A sample can be prepared that is enriched in a particular isotope, then administered into living things. If the isotopes are radioactive, they can be detected by the radiation they give off. If they are stable, they can be detected from their mass by analysis in a machine called a mass spectrometer.

The presence of different isotopes of the atoms of each element is taken into account when calculating the average mass of one atom of the element, known as its 'relative atomic mass'. For example, there are two isotopes of chlorine, of mass 35 and mass 37. 75% of chlorine atoms are chlorine-35 while 25% are chlorine-37. Hence the average mass of a chlorine atom is $(35 \times 0.75) + (37 \times 0.25) = 35.5$.

see also...

Atoms, Carbon Dating, Elements, Radioactivity, Sub-atomic Particles

Lavoisier, Antoine

The French nobleman Antoine Lavoisier (1743–1794) is known as the 'Father of Modern Chemistry'. He performed experiments that placed emphasis on meticulously measuring the changes that accompany chemical reactions. His most famous discovery was that the mass of chemicals at the start of a reaction equals the mass at the end. This is the Law of Conservation of Mass. For example, when iron reacts with oxygen to form iron oxide, the mass of the iron oxide is found to be equal to the combined mass of the original iron and oxygen. This seems very obvious to chemists today, because we know that the atoms present at the start of a reaction must also be present at the end of the reaction. That is the basic rule we use when balancing chemical equations. Lavoisier, however, was working before John Dalton had developed his atomic theory. He did not have the knowledge required to think about reactions in the way that modern chemists do, but his observations laid the foundations for that knowledge to develop.

There is an interesting twist to this story, however. We now know that the Law of Conservation of Mass is not really true. The work of Albert Einstein has revealed that 'energy has mass', so the masses of reactants and products in chemical reactions *do* change, depending on how much energy is lost or gained as the reaction proceeds. The mass change (m), associated with an energy change (E) is quantified by Einstein's equation: $E = mc^2$. 'c^2' is the speed of light squared, which is so large it means that tiny amounts of mass correspond to large amounts of energy. No chemistry laboratory balance will ever detect the incredibly small mass changes due to the energy lost or gained during chemical reactions.

So, the Law of Conservation of Mass remains a valid observation for practical chemistry. This subtlety demonstrates that 'laws' and theories that we know are not precisely true can nevertheless prove very useful to us because they remain true in many practical circumstances. This is a situation that occurs quite often in science.

see also...

Dalton, John, Energy, Equations

Lewis Structures

When representing the structures of molecules it is often important to keep track of the way in which the outer electrons of the various atoms are distributed. This helps us understand how the electrons form the covalent bonds that hold molecules together, and lets us check that all the electrons are accounted for. The most common way of doing this is by drawing Lewis structures (also known as Lewis electron-dot structures) in which dots or crosses are used to represent electrons and lines are used to represent covalent bonds.

The formation of a hydrogen molecule (H_2) can be represented using Lewis structures as follows:

H• + •H ⟶ H:H or H – H

A dot represents each electron, and the Lewis structure of the molecule indicates that the two electrons are shared between the two hydrogen atoms. Each shared pair of electrons constitutes a covalent bond, which can be represented by a simple line.

The Lewis structure of water is as follows:

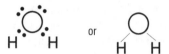

This shows two pairs of outer electrons that are not involved in bonding (known as 'lone pairs') in addition to the two pairs that form the covalent bonds between the oxygen atom and hydrogen atoms.

Double bonds and triple bonds can also be represented by Lewis structures, illustrated by the structures of nitrogen (N_2) and carbon dioxide (CO_2) below:

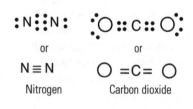

see also...

Covalent Bonds, Electron Configuration

Metals

ivilisation has been built around metals. We talk of the Iron Age and Bronze Age, which gave us tools, utensils and weapons for both hunting and war. Silver and gold have long been prized for jewellery. The industrial revolution saw machinery built from metal being used to achieve many previously impossible, or extremely laborious, tasks. Metals were used to build ships, trains, trucks, cars and aircraft. Nowadays we have moved far beyond the exploitation of the traditionally used metals such as iron, copper, tin, zinc, silver and gold. We now use many more exotic metals such as aluminium, titanium, magnesium, tungsten and plutonium. The uses of metals range from the sublime to the almost ridiculous. Metals formed the spacecraft that took men to the moon, and unmanned probes beyond the solar system; but even golf balls are sold with boasts of the titanium and magnesium they contain.

The vast majority of elements are metals, yet there is no single definition for what a metal actually is. Metals are defined by their possession of a range of characteristics, including their shiny ('lustrous') nature when solid; their malleability, allowing them to be beaten into various shapes; and their ability to conduct electricity. These properties all stem from perhaps their most characteristic property, which is the type of chemical bonding that holds the structure of a metal together. Solid metals consist of positive ions held together by a 'sea' of the detached outer electrons of the metal atoms. This situation is known as 'metallic bonding', and explains why metals conduct electricity (due to the mobile electrons) and are malleable (due to the rather fluid and changeable structure within them).

Many of the most useful substances that we think of as metals are actually homogeneous mixtures of two or more metallic elements, known as 'alloys'. Bronze is an alloy of copper and tin. Steel, our most commonly used 'metal' is not purely metal at all, since it is an alloy of iron with small quantities of carbon (a non-metal) and other metallic elements.

see also...

Atoms, Electron Configuration, Elements

Minerals

To a geologist, minerals are the naturally occurring elements and compounds of the Earth's crust. Most rocks, ores and natural crystals are minerals, such as silicon dioxide (SiO_2 – which forms quartz and sand), calcium carbonate ($CaCO_3$ – which forms limestone, chalk and marble) and iron oxide (Fe_2O_3 – which is iron ore). Minerals are inorganic materials, so do not include the organic compounds of plant and animal life. Soil is a mixture of minerals (rock, stone, grit and dissolved mineral ions) and organic materials (such as plant fibre, rotting animal and plant material, bacteria and fungi).

In nutrition, the term minerals refers to a variety of elements that are essential for health, but are not supplied by the major nutrients (proteins, carbohydrates and fats). The 'micronutrient minerals' are mineral elements we need in tiny quantities. These include iron, zinc, copper, iodine, selenium, manganese, molybdenum, cobalt and chromium. A healthy diet may contain just one milligram of copper per day, for example, but that tiny amount, on average, has to be there or we will become ill. There are also seven 'macronutrient minerals' which we need in larger amounts. These are calcium, phosphorus, potassium, sulphur, sodium, chlorine and magnesium. Calcium accounts for about 2% of body weight. It is well known as an important component of bones and teeth, but also plays vital roles in the chemistry of every living cell. We need to consume about one gram of calcium each day. That is a tiny amount compared to the mass of food we eat overall, but is substantially more than the required amounts of the micronutrient minerals.

'Mineral water' is water that has percolated through rocks and underground waterways, and therefore contains relatively high quantities of dissolved minerals, especially calcium carbonate, magnesium sulphate, potassium sulphate and sodium sulphate. Carbonated mineral water contains dissolved carbon dioxide gas, which can be released from calcium carbonate, producing the water's 'fizz'.

see also...
Elements

Molecular Bonding Theories

Molecules are held together by electrons being shared between the nuclei of the atoms involved. The forces of attraction between the negative electrons and the positive nuclei hold the structure together. Chemists describe the forces of attraction between the atoms as chemical 'bonds' but they use two different theories or 'models' of bonding to describe these bonds.

Valence bond theory considers a covalent bond to be formed when two electrons occupy a bonding orbital that is localized around two neighbouring nuclei in a molecule. The bonding orbital occupied by the shared electrons is created by the overlap of two orbitals from the atoms that become bonded together. The bonds are regarded as involving only electrons from the outer energy level of each atom. The inner electrons are assumed to remain localized around the nuclei of the atoms they originally came from.

Molecular orbital theory is an alternative model of bonding, which provides a better explanation for certain aspects of chemical behaviour. In molecular orbital theory all the electron orbitals on every atom of a molecule are mathematically combined. This yields a set of 'molecular orbitals' which cover the entire structure of the molecule. Some of these orbitals are 'bonding orbitals', meaning that electrons occupying them will hold the molecule together. Other molecular orbitals are 'anti-bonding', meaning electrons occupying them will encourage the molecule to fall apart. A few molecular orbitals are 'non-bonding', meaning they neither assist nor counteract the forces holding a molecule together. Electrons generally occupy the lowest energy orbitals available. If that leads to more electrons being in bonding orbitals than in antibonding orbitals, then a stable molecule can be formed.

Valence bond and molecular orbital theories are two different representations of the same reality.

see also...

Covalent Bonds, Electron Orbitals, Ionic Bonds

Molecular Electronics

Molecular electronics is the name given to a broad field of research devoted to using individual molecules, or small groups of molecules, to replace the conductors, transistors, diodes and other electrical components of present-day microelectronics. Molecular electronics promises to take the miniaturisation of electronics into a new, and even tinier realm. Its components may be less than one-hundredth the size of the already very small semiconductor-based components of microelectronics today. This is one aspect of 'nano-technology' – building technological components and machines on the smallest possible scale.

The likely main components of molecular electronic circuits will be organic (carbon-based) molecules, including some very similar to the naturally occurring components of living things. This is a field in its early stages of development, but already there have been significant achievements. The Nobel Prize in Chemistry for 2000 was awarded to researchers who have developed electrically conducting polymers – chain-like organic compounds that might act as the molecular 'wires' of molecular-scale circuits. Other researchers have devised various kinds of molecular switches. These contain individual molecules that can flip between two or more states, which could be used as the basis of information storage or to change the paths electrons follow through molecular electronic circuits.

It is too early to know if the potential people see in molecular electronics will ever be realised, or exactly what applications it will lead to. If even the more modest plans work out, palmtop computers many thousands of times more powerful that a modern-day desktop could be feasible. More adventurously, electrical components may be developed that essentially assemble themselves, in a similar way to the self-assembling membranes, organelles and cells of living things. We could also devise circuits that mimic the chemistry of photosynthesis, trapping free and clean energy from the sun.

see also...

Molecules

Molecules

Molecules are particles composed of two or more atoms bonded together by the sharing of electrons. We can make sense of that definition by looking at the simplest molecules of all – hydrogen molecules (H_2) composed of two bonded hydrogen atoms.

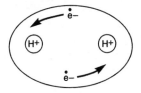

A hydrogen molecule (H_2)

A hydrogen molecule is formed from two hydrogen atoms, each composed of a nucleus (H^+), surrounded by an electron (e^-). In the hydrogen molecule the two electrons derived from each hydrogen atom move in an orbital that surrounds both nuclei. We can view this as a situation in which the force of attraction between the two nuclei and the electrons allows the electrons to serve as a kind of 'glue' holding the nuclei together. The orbitals occupied by electrons within molecules are known as 'molecular orbitals'.

The bonds holding molecules together are called 'covalent bonds', and can be sub-divided into pure covalent bonds, between identical atoms, and 'polar covalent bonds' between atoms with significantly different charges on their nuclei. One pair of bonding electrons being shared between two atoms is defined as a single 'bond', and is represented by a line in structural formulas. So H–H denotes two hydrogen atoms joined by a single covalent bond. Using this symbolism the structures of some other common molecules are shown below.

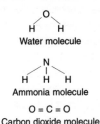

Water molecule

Ammonia molecule

$$O = C = O$$
Carbon dioxide molecule

Molecules come in all sorts of shapes and sizes, ranging from hydrogen, all the way up to molecules of DNA, which contain many millions of atoms bonded together.

see also...

Atoms, Covalent Bonds, Deoxyribonucleic Acid (DNA), Electron Orbitals, Polar Covalency

Moles

Chemists quantify the amounts of substances using the concept of the mole. One mole of any chemical is the amount that contains 6.02×10^{23} atoms, molecules or formula units, depending on the substance concerned. The number 6.02×10^{23} is known as the Avogadro Constant, or Avogadro's Number. It has been called 'The Chemist's Dozen', since it is the basic counting unit of chemistry. The mole is the internationally accepted SI (for *Systeme International*) unit for 'amount of substance'. It is formally defined as 'the amount of substance that contains as many specified entities as there are carbon atoms in exactly 12 g of the carbon-12 isotope'. Since that number is 6.02×10^{23}, we can get by with the much simpler practical definition that 'one mole of "things" = 6.02×10^{23} "things".'

One mole of copper metal, composed of copper atoms (Cu), will contain 6.02×10^{23} copper atoms. One mole of water, composed of water molecules (H_2O), will contain 6.02×10^{23} water molecules. One mole of sodium chloride, an ionic substance composed of sodium and chloride ions (NaCl), will contain 6.02×10^{23} NaCl formula units.

The mass of any substance that contains exactly one mole of the substance is known as its 'molar mass', with units of grams per mole ($g \, mol^{-1}$). For example, the molar mass of water is $18 \, g \, mol^{-1}$, found by adding the relative atomic masses of the oxygen and two hydrogen atoms present in a water molecule $(16 + [2 \times 1])$.

The volume of any substance that contains exactly one mole of the substance is known as its 'molar volume'. Most gases, for example, have a molar volume very close to 22.4 litres at room temperature and atmospheric pressure.

The concentration of a solution, meaning the amount of any dissolved substance present in a given volume of a solution, is measured in moles per litre ($mol \, l^{-1}$). This value is known as a solution's 'molarity'.

see also...

Atoms, Isotopes, Molecules

Nitrogen Cycle

Air is mainly nitrogen (80%) in the form of N_2 molecules which are very unreactive because the atoms are held together by a strong triple covalent bond ($N\equiv N$). This unreactivity makes nitrogen a relatively ignored component of air. Nitrogen is, however, an essential element for plants and animals, needed to make amino acids, proteins, DNA and many other crucial chemicals of life. The continual cycling of nitrogen atoms within the environment, including its movement among the atmosphere, plants and animals, is called the nitrogen cycle. The nitrogen cycle is of huge environmental and industrial relevance, and human activity disturbs it in various significant ways.

Animals and plants cannot obtain supplies of nitrogen directly from the nitrogen of the air. Bacteria, however, can convert nitrogen into compounds such as nitrates (containing the NO_3^- ion), which plants can absorb through their roots. Animals get their supplies by eating plants. The conversion of nitrogen into nitrogen-containing compounds is called 'nitrogen fixation'. The 'leguminous plants', such as peas and clover, have nitrogen-fixing bacteria living in their root nodules, hence they behave as if they were nitrogen-fixing plants, and can be grown and then ploughed in to raise the fertility of soils. Nitrogen fixation also occurs naturally during lightning storms, which provide sufficient energy to make nitrogen react with oxygen in the air. Nitrogen returns to the air through the activities of 'denitrifying' microbes, which release nitrogen gas from nitrogen-containing compounds during the decomposition of organic matter such as dead plants and animals.

The chemical industry 'fixes' huge amounts of nitrogen during the manufacture of nitrogen-containing fertilizers. This artificial nitrogen fixation begins with the Haber Process in which nitrogen and hydrogen are combined to form ammonia (NH_3). The ammonia can be used as a fertilizer directly, or can be converted into compounds such as ammonium nitrate and ammonium phosphate.

see also...
Carbon Cycle, Fertilizers

Noble Gases

The elements in Group 8A (sometimes called Group 0), at the extreme right of the Periodic Table, are known as the noble gases, or inert gases. The six noble gases are helium, neon, argon, krypton, xenon and radon.

These elements are extremely unreactive due to their uniquely stable electron configurations. This lack of reactivity means they occur as free atoms, always in the gaseous state at everyday temperatures and pressures. They do not generally combine with other elements to form compounds. The name, 'noble gases' stems from a comparison of their unwillingness to combine with other elements to the stand-offishness of the 'nobility', unwilling to mix with common people.

The stability of the noble gases reveals one of the most significant chemical principles underpinning the reactivity of other elements. All the noble gases have a 'stable octet' of eight outer electrons, apart from helium, which has only two electrons. Helium, however, does possess a completely full outer energy level (or 'outer shell') of electrons. When elements other than the noble gases react to form compounds, the electron configurations of their atoms often change to become like the very stable electron configurations of noble gas atoms.

Elementary chemistry texts often express this important principle by stating that atoms react 'in order to attain a noble gas electron configuration'. In reality, atoms have no sense of purpose, since they are mere assemblies of sub-atomic particles. The correct explanation is that during the course of chemical reactions the electron configurations of other atoms can settle into, and become trapped in, the stable electron configurations found in the noble gases. Atoms can attain a noble gas electron configuration either through the transfer of electrons, leading to the formation of ions, or by the sharing of electrons in covalent bonds.

---

see also...

Covalent Bonds, Electron Configuration, Ionic Bonds, Periodic Table

Nucleosynthesis

Where have the atoms of all the elements come from? The essential answer is that the simplest atoms, those of hydrogen, formed from sub-atomic particles in the aftermath of the Big Bang. All the larger atoms were then assembled by nuclear fusion processes in which the nuclei of small atoms combined to form those of larger atoms. The formation of the nuclei of new atoms from existing ones is called nucleosynthesis.

Simple examples of nucleosynthesis occur in the Sun and release the energy of sunlight. Inside the Sun, the nuclei of hydrogen atoms combine to form the nuclei of helium atoms. The Sun is currently 71% hydrogen and 27% helium, so it has sufficient hydrogen 'fuel' to burn for a long time to come. Smaller amounts of other elements also form in the Sun and other stars. For example, helium nuclei can fuse to form lithium or beryllium nuclei. Fusion between beryllium and helium can form carbon. Fusion of carbon with hydrogen yields nitrogen.

Eventually, many stars run out of fuel, contract, then explode in a spectacular supernova explosion. This scatters a star's harvest of atoms into space, and the explosion also powers additional nucleosynthesis processes that create other kinds of atoms. Some of the atoms created are unstable, and can split into two different nuclei. The precise details are complex and in some cases not fully understood. The basic concept, however, is wonderfully simple: the nuclei of big atoms are made from those of smaller atoms when they combine to form different accumulations of the protons and neutrons that all nuclei are made of. Once new atomic nuclei are formed, they can combine with the appropriate number of electrons, to form electrically neutral atoms, whenever the prevailing conditions are sufficiently low in energy to allow this to happen.

In recent years we have learned how to perform nucleosynthesis for ourselves, to create various elements that do not occur naturally on Earth.

see also...

Elements, Periodic Table, Sub-atomic Particles

Organic Chemistry

The chemistry of almost all carbon-containing compounds is known as organic chemistry. The origin of the phrase is the fact that the key chemicals of living *organisms* are based on chains or rings of carbon atoms, with other atoms attached as side groups or incorporated within the chains or rings. The fossil fuels, coal, oil and natural gas, are carbon-containing chemicals derived from ancient living things, so they are also part of organic chemistry. The coverage of the term is broadened even further to include all the carbon-based chemicals we make from coal, oil or natural gas. So, organic chemicals include all the synthetic plastics, fibres, medicines, pigments and adhesives produced by the petrochemical industry. The main carbon-containing compounds that are not regarded as organic chemicals are carbon dioxide (CO_2) and carbon-containing rocks such as limestone and marble (both forms of calcium carbonate, $CaCO_3$). These are referred to as inorganic carbon compounds. Inorganic chemicals comprise all chemicals apart from those carbon-containing ones we call organic.

Organic chemistry is a huge and very influential branch of chemistry because it covers the chemistry of life, of most medicines, and of the most common synthetic materials such as plastics. One reason for the vast range of organic chemicals is the ability of carbon atoms to bond to themselves to form chains of any length, a wide variety of ring structures, and to also bond to atoms of hydrogen, oxygen, sulphur, nitrogen, phosphorus, fluorine, chlorine, bromine and iodine. This is related to the fact that carbon has an intermediate electronegativity value, meaning an intermediate 'pulling power' for the shared electrons within covalent bonds, so it can form stable bonds with many different types of atom. Many chemists believe any life elsewhere in the universe will also be based on organic compounds similar to our own.

see also...

Alcohols, Biochemistry, Carbohydrates, Deoxyribonucleic Acid (DNA), Electronegativity, Esters, Fats and Oils, Petrochemistry, Proteins, Ribonucleic Acid (RNA)

Oxidation

Oxidation is one of many chemical terms that can confuse the unwary, because it doesn't mean what it sounds as though it should mean. The original meaning of oxidation was 'reaction with oxygen', which is clear enough. It was also used for the loss of hydrogen, since the hydrogen was generally taken away by reaction with oxygen to form water. The meaning of oxidation became more general, however, as more was learned about what happens as chemicals react with oxygen. In its modern sense, oxidation occurs when a chemical loses electrons and/or reacts with oxygen, and/or loses hydrogen. Why has this broadening of meaning occurred? If we explore what happens to metals, for example, that react with oxygen, we find that the metal atoms generally lose electrons to form positive ions, while the oxygen atoms acquire the electrons and form negative oxide ions. For example:

$$2Ca + O_2 \rightarrow 2\,[(Ca^{2+})(O^{2-})]$$

This is essentially an *electron transfer* reaction in which calcium loses electrons, while oxygen gains electrons. So the loss of electrons became the basic definition of oxidation, even when oxygen is not involved. So if calcium reacts with chlorine (Cl_2) instead of oxygen, the calcium is still 'oxidized' because it loses electrons to the chlorine:

$$Ca + Cl_2 \rightarrow (Ca^{2+})(Cl^-)_2$$

The opposite of oxidation, the *gain of electrons*, is known as 'reduction'. Reduction and oxidation are the two halves of what are known as 'redox' reactions. Redox reactions can be set out in equation schemes showing the oxidation and reduction processes, and the overall redox reaction as follows:

$Ca \rightarrow Ca^{2+} + 2e^-$	Oxidation
$O_2 + 4e^- \rightarrow 2O^{2-}$	Reduction
$2Ca + O_2 \rightarrow 2CaO$	Redox

The ion-electron equation for the oxidation of calcium (top line) must be multiplied by two when combining both half-equations, since each oxygen molecule reacts with four electrons.

see also...

Electronegativity, Polar Covalency, Reduction

Ozone

Ozone (O_3) is a rare form of oxygen that has three oxygen atoms per molecule, instead of the more common two. One-fifth of the air is composed of oxygen in its normal 'diatomic' form (O_2), which must be continuously absorbed into our blood to keep us alive. Tiny amounts of this oxygen are converted into ozone by ultra-violet (UV) radiation from the sun. The formation of ozone occurs in a two-step process as follows:

$$O_2 + UV \ Energy \rightarrow 2O$$
$$2O + 2O_2 \rightarrow 2O_3$$

The ozone, however, is normally destroyed as fast as it is produced by the following decomposition steps:

$$O_3 + UV \ Energy \rightarrow O_2 + O$$
$$O + O_3 \rightarrow 2O_2$$

Overall, UV rays from the sun power a continual cycle of ozone production and decomposition which maintains a very low concentration of ozone in the atmosphere. The ozone concentration is highest in the 'ozone layer' about 10 – 40 kilometres above sea level; but even the ozone layer contains only tiny amounts of O_3 compared with O_2. This ozone is very significant to us, however, because it absorbs significant amounts of ultra violet radiation. This protects ourselves and other living things from the harmful effects of that radiation, which include the ability to induce mutations in DNA, leading to cancer.

The depletion of the ozone layer, especially above the poles, by chemical pollutants is one of the best publicized environmental problems caused by our exploitation of chemistry. The main 'villains' are chemicals called 'chlorofluorocarbons' (CFCs) used as refrigerants, aerosol propellants and industrial solvents. The CFCs, and other ozone-depleting pollutants, react with ozone and reduce its steady-state concentrations. There are now promising signs that the phasing out of CFCs, and their replacement by 'ozone-friendly' alternatives, may allow the 'ozone hole' to repair itself within about 50 years.

Although ozone is a 'good thing' up in the stratosphere, it is a hazardous component of pollution and photochemical smog at ground level.

see also...

Green Chemistry, Smog

Particulates

Very tiny pieces of material suspended in the atmosphere are known as particulates. They can be solid, liquid, or a mixture of both. Particulates are key parts of fumes, dust, and smog. They are important components of polluted air, responsible for problems including irritation of the eyes, throat and nose, serious respiratory conditions and perhaps cancer. Many particulates are natural, but some of the most troublesome result from human activity. Natural particulates include lumps and droplets of material dispersed into the atmosphere from the oceans, dusty dry land, ejected from volcanoes, or released in the smoke of forest fires. These are supplemented by particulates produced during the combustion of petrol, diesel, oil, cigarettes, and released in smokestacks from a range of industrial processes. Some particulates are created in the atmosphere from the reaction of gases and liquids already present in the air.

As soon as any particulate is formed it can serve as a chemically active surface on which a wide variety of chemical reactions can occur. These can change the chemical nature of the particulate and alter its size and structure. The particulates in urban air of most relevance to us are categorized into 'coarse' and 'fine' forms. The fine forms tend to be acidic, chemically reactive, and form as a result of combustion in vehicle engines and industrial processes. They range from around 0.1 to 1 millionth of a meter across (0.1 to 1 micrometres). The coarse particulates are generally produced by natural processes such as the dispersal of dust from soil. They range upwards from 1 to 100 micrometres across.

The smaller particulates generally pose the greatest health risks, irritating the mucous membranes lining our airways and seriously aggravating any pre-existing respiratory problems such as asthma. Studies suggest they may have 'mutagenic' effects, triggering events that cause the changes in DNA that can lead to the onset of cancer. The particulate content of the air is now routinely monitored.

see also...

Smog

Periodic Table

The Periodic Table of the Elements lists all the chemical elements – substances containing only one type of atom. Thus, the Periodic Table also lists all the types of atoms. All substances are either elements or compounds, and since compounds are composed of different elements bonded together the Periodic Table displays the building blocks of every chemical – whether natural or human-made.

The Periodic Table is much more than a simple list of elements. Its structure embodies crucial information about the chemical properties of the elements and the structure of their atoms. Vertical columns in the table are called 'Groups' and elements in any particular group share important chemical characteristics. The elements in Group 1A, for example, are reactive metals that form alkalis on reaction with water. The elements in Group 8A (also called Group 0) are unreactive gases. The fact that certain chemical properties recur in a periodic manner, as we move through the table, is what gives the 'periodic' table its name. Atoms of elements in the same main group (from 1A to 8A) have the same number of electrons in their outer energy level. The horizontal rows in the table are called 'Periods'. The Period number indicates how many occupied electron energy levels there are in the atoms concerned. Thus elements in Period 1 have their electrons in the first energy level; elements in Period 2 have electrons in both first and second energy levels, and so on. The different 'blocks' of the Periodic Table contain atoms whose outer electrons are in particular types of orbital. This gives us the s-block, p-block, d-block and f-block, whose atoms have outer electrons in s, p, d and f orbitals respectively.

The modern Periodic Table grew from the work of several great scientists, most notably the Russian Dimitri Mendeleev, who is known as the 'Father of the Periodic Table'.

An internet Periodic Table is available at http://www.webelements.com

see also...

Atoms, Compounds, Electron Configuration, Electron Orbitals, Elements

Petrochemistry

Crude oil, or 'petroleum', is our most versatile chemical resource. It is converted by the petrochemical industry into a huge variety of fuels, plastics, fabrics, pharmaceuticals and other industrial and consumer products. Crude oil is essentially a complex mixture of hydrocarbons, composed of chains or rings of bonded carbon atoms with many hydrogen atoms attached. It is derived from ancient plant and animal life, so when used as a fuel it is known as 'fossil fuel'. The first step in processing crude oil is to separate it into different 'fractions' containing hydrocarbons with broadly similar boiling points. This is achieved by fractional distillation. The fractions with the largest molecules, containing more than 18 carbon atoms per molecule, form a thick residue used to make asphalt and tar. Fractions with between 12 and 18 carbon atoms are used for lubricating oils and fuel oil. Those with around six to 10 carbon atoms per molecule are converted into petrol, diesel and kerosene. The fractions with the smallest molecules are composed of gases such as butane, propane, ethane and methane.

Most of the chemicals we purify from crude oil are burned as fuels, the most wasteful thing we could do with them. These hydrocarbons are a versatile chemical resource used to make some of our most valuable materials and pharmaceuticals. For example, the very simple hydrocarbon ethene (C_2H_2) is the primary raw material for a vast range of chemical conversions. These produce the plastics polyethene (polythene), polystyrene, polyvinyl chloride (PVC), and many other useful products.

Three of the most fundamental processes of petrochemistry are 'cracking', which breaks large hydrocarbons into smaller ones; 'reforming', which changes the bonding arrangement of hydrocarbons; and 'polymerisation', which links many small molecules into large polymers. Petrochemistry also relies on various reactions that add other functional groups (OH, Cl, Br, COOH, etc.) to a basic hydrocarbon framework.

see also...

Functional Groups, Hydrocarbons, Polymers

pH

The pH scale quantifies the concentration of hydrogen ions (H^+) in any solution. However, pH values actually reveal both the concentration of hydrogen ions (characteristic of acids) and of hydroxide ions (OH^-) characteristic of alkalis. Each pH value indicates whether a solution is acidic, alkaline or neutral. The pH scale is familiar to many gardeners, who may need to determine the pH of the water in their soil. The pH values of lakes and rivers are often quoted to indicate the extent to which they have been affected by 'acid rain' and other sources of pollution.

A solution with pH 7 is 'neutral' in the sense that it is neither acidic nor alkaline. A neutral solution contains an equal number of H^+ ions and OH^- ions. Acids have pH values less than 7, while alkalis have pH values greater than 7. So acidity increases as pH values decrease.

The term 'pH' is shorthand for the negative logarithm, to the base 10, of the hydrogen ion concentration. This relationship can be summarised:

$$pH = -\log_{10} [H^+]$$

The following list of hydrogen ion concentrations and pH values makes the link between H^+ concentration and pH clear:

H^+ concentration	pH
0.1 moles per litre	1
0.01 moles per litre	2
0.000 000 1 moles per litre	7
0.000 000 000 000 1 moles per litre	13
0.000 000 000 000 01 moles per litre	14

So, in whole number pH values, the pH indicates how many places the decimal point in the concentration value is moved to the left of 1. The pH scale is a logarithmic scale. This means that each change in pH by one unit corresponds to a ten-fold change in hydrogen ion concentration. For example, lake water with a pH value of 4 contains 10 times the amount of hydrogen ions in a given volume as lake water with a pH of 5.

The concentration of OH^- ions can be calculated from any pH value because of this fixed relationship:
H^+ concentration $\times OH^-$ concentration
$$= 1 \times 10^{-14} \text{ moles}^2 1^{-2}$$

see also...
Acid Rain, Acids, Bases

Photosynthesis

hen sunlight hits the green parts of a plant, some of it is absorbed and powers the conversion of water and carbon dioxide into carbohydrates, with oxygen gas being released as a by-product. This is the process known as photosynthesis, which allows plants to use sunlight as their sole source of energy. The chemistry of photosynthesis can be summarised as follows:

$$6H_2O + 6CO_2 \rightarrow C_6H_{12}O_6 + 6O_2$$

water   carbon   glucose oxygen
      dioxide

The carbohydrate, glucose, can serve as a raw material used to manufacture other chemicals needed by the plant. It can also be used as a source of chemical energy to power other energy-requiring reactions in the plant. When used as a source of energy, the glucose is oxidised back to carbon dioxide and water in a reversal, overall, of photosynthesis. Some of the energy released serves to manufacture ATP (adenosine triphosphate) – the central chemical 'energy currency' that powers most energy-requiring reactions of life. The oxidation of glucose in this way is called respiration, and is the main process that enables humans and other animals to obtain energy by the oxidation of food molecules (such as glucose). Photosynthesis and respiration are two arms of a solar-powered cycle that sustains life on earth.

Photosynthesis is a very complex chemical process, achieving the reaction shown above through a long series of steps. The primary event, however, is the absorption of light by green chlorophyll pigments in plant cells. The light drives electrons out of chlorophyll molecules and initiates a cascade of events that converts the energy of the ejected electrons into chemical energy stored within the chemicals of a plant. If we could efficiently mimic this process within artificial chemical systems that could generate electricity, we would have a free, clean and enduring source of energy. Many prototype systems have been developed, and as their efficiency gradually improves they offer great hope as a primary energy source for the not too distant future.

see also...

*Adenosine Triphosphate (ATP),
Carbohydrates, Photovoltaics*

Photovoltaics

The sun floods the Earth with much more energy than we need to power all our energy-hungry activities, but we make relatively little use of this free and clean energy. Photovoltaics is the direct conversion of light energy into electrical energy. We already use photovoltaic 'solar cells' to power many appliances, including watches, calculators, lighthouses and spacecraft. The real promise of the technology, however, will be fulfilled when it powers cars, trucks, trains, homes, offices and factories, and allows us to escape from our dependence on polluting and limited fossil fuels.

The key chemical event within photovoltaic systems is the absorption of light by a chemical, accompanied by the ejection of electrons to produce an electric current. This is also the key event in photosynthesis, the natural form of photovoltaics occurring in plants. Instead of the chlorophyll molecules in plant systems, artificial photovoltaics largely relies on solid semiconductors to absorb light and release electrons. Silicon and gallium arsenide are two of the most common semiconducting materials used to fabricate solar cells. One of the most interesting areas of research is aimed at producing photovoltaic systems that are transparent to light. This would allow the huge surface areas available in windows of buildings to double up as electricity generators powering the activities within. In some prototypes, dye molecules containing ruthenium ions are used to absorb some of the visible light falling on them. The dye molecules are coated on nanocrystals of titanium dioxide (TiO_2), which is a semiconductor. The energy of the light ejects electrons from the dye molecules, the titanium dioxide crystals conduct these electrons away from the dye and transfer them into the external electrical circuit. Iodide ions trapped within the solar cell supply electrons to replace those ejected from the dye molecules; and the electrons returning from the external circuit replace these lost electrons. With further research the efficiency of such systems can be expected to steadily increase.

see also...

Photosynthesis, Semiconductors

Polar Covalency

Molecules are held together by covalent bonds. These are due to the forces of attraction among atomic nuclei and electrons when some electrons become shared between two or more atoms. In many bonds, however, the sharing is unequal, in that the shared electrons are drawn much more strongly towards one of the bonded atoms than another. This causes the bond to be 'polarized' into one pole that has an excess of negative charge and another pole with a deficiency of negative charge. The pole with a slight negative charge is known as the 'delta minus' ($\delta-$) pole, while the other pole, with a slight positive charge, is the 'delta plus' ($\delta+$) pole. Any bond with $\delta-$ and $\delta+$ poles is called a *polar covalent bond*. The $\delta-$ pole is found on the atom with the highest electronegativity, i.e. greatest attraction for electrons, while the $\delta+$ pole is on the atom with the lowest electronegativity. Common examples of polar covalent bonds are the O–H, Cl–H and O=C bonds. In each of these cases the atom on the left has the greater electronegativity, and so carries the $\delta-$ pole.

Polar covalent bonds have a great influence on both the reactivity and the physical properties of chemicals containing them. The $\delta-$ and $\delta+$ poles act as regions which oppositely charged groups on other chemicals can be drawn towards in ways that initiate the redistribution of electrons and bonds involved in chemical reactions. A huge variety of reactions are initiated and guided by the charged regions at either ends of polar covalent bonds. Polar covalency also allows weak non-covalent interactions to be set up among molecules, such as the so-called 'hydrogen bonds' among water molecules.

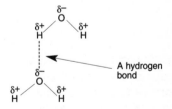

A hydrogen bond

see also...

Covalent Bonds, Electronegativity, Intermolecular Forces, Molecular Orbitals, Polar Molecules

Polar Molecules

Many molecules have an asymmetric distribution of electric charge, causing one end or side of the molecule to have a partial positive charge ($\delta+$) and another end or side to have a partial negative charge ($\delta-$). Such molecules are called polar molecules, since they have $\delta+$ and $\delta-$ 'poles', in the same way as the Earth has North and South poles. The asymmetric distribution of charge is due to the presence of polar covalent bonds, arranged in an asymmetric manner. A water molecule, for example, has the following structure:

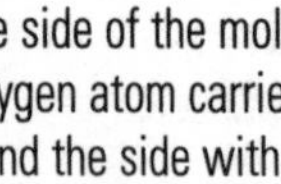

Water – a polar molecule

Water molecules are polar overall because the side of the molecule with the oxygen atom carries a $\delta-$ charge and the side with the hydrogen atom carries a $\delta+$ charge.

The tetrachloromethane molecule shown below also contains polar covalent bonds but it is not a polar molecule overall because these bonds are 'symmetrically opposed' and so cancel out overall. In other words, the molecule does not have any particular end or side that is more negatively or positively charged than any other end or side.

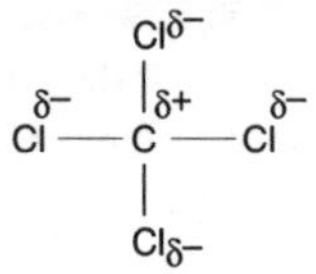

Tetrachloromethane – a non-polar molecule

Polar molecules tend to mix very easily with other polar molecules, because of the interactions between their oppositely charged regions. This is a crucial factor that governs which molecules are soluble in water. Polar molecules, such as sugars, dissolve readily in water. Non-polar molecules, such as hydrocarbons, fats and oils, do not dissolve in water. Some molecules have polar regions bonded to non-polar regions, such as the soap and detergent molecules that allow greases and oils to mix with water, although not properly dissolve.

see also...

Polar Covalency, Soaps and Detergents

84

Polymers

When many identical or structurally similar small molecules bond together to form extremely large molecules, we describe the large molecules as polymers (from the Greek **poly** = 'many' and **meros** = 'parts'). The small molecules that link up to form a polymer are called *monomers* (**mono** = one). Both Nature and industrial chemists make great use of polymers to build the materials of the world. In Nature, proteins, DNA, RNA and complex carbohydrates such as starch and cellulose are all polymers. Synthetic polymers include all the plastics and synthetic fibres and resins of modern life. Familiar terms such as nylon, polythene, polypropylene, polyester, polytetrafluoroethylene (Teflon), and Kevlar are all names of various synthetic polymers. The basic raw material we use for making synthetic polymers is crude oil, which supplies the monomers required, or small molecules that we can transform into the monomers required. In the not too distant future, as oil becomes scarce, we may increasingly make synthetic polymers out of renewable sources of organic chemicals, namely suitable crop plants.

When monomers simply add together, with no side products, an 'addition polymer' is formed. When monomers add in reactions that release small molecules such as water as a by-product of forming each bond, a 'condensation polymer' is formed. The polymers can be linear polymers or branched polymers, depending on whether or not they contain cross-linkages between neighbouring parts of the polymer chain. They may also be 'thermoplastic' polymers, which soften on heating, or 'thermosetting' ones that harden when heated. Many synthetic polymers do not readily degrade (break down) when discarded, unlike the natural polymers in wood, for example. Environmental concerns have led to the development of some biodegradable synthetic polymers; and also of techniques to recycle non-biodegradable types.

see also...

Addition Reactions, Carbohydrates, Condensation and Hydrolysis Reactions, Deoxyribonucleic Acid (DNA), Esters, Molecular Electronics, Proteins, Ribonucleic Acid (RNA)

Proteins

Every living thing is assembled, maintained, and substantially composed of the chemicals we call proteins. Protein molecules form the enzymes that catalyse almost all the chemical reactions of life. Proteins form many of the structural materials within living cells and tissues. Skin, bone and cartilage are based on a framework of protein, and every cell has an internal 'cytoskeleton' made of protein tubules. Our ability to move depends on contractile fibres made of protein molecules that slide past one another. Protein molecules act as chemical transporters, binding to chemicals in one part of a living thing and transporting them to other parts. The protein, haemoglobin, for example, whisks oxygen around our bodies in this way. Proteins also carry chemical 'messages' from place to

embedded in cell membranes act as 'gates' and 'pumps', controlling what flows into and out of cells. The nervous activity that allows you to read this, and think about it, is sustained by protein gates and pumps opening and closing in your nerve cell membranes. Proteins control the activities of other chemicals of life by binding to them and either activating them or inhibiting them.

Every protein is a 'polyamide' formed when various amino acid molecules become linked into a protein ('polypeptide') chain. Each chain folds into a highly specific structure, required for it to perform its chemical activities.

The repeating structure of a protein chain. Twenty different 'R' groups occur, in varying sequences, in molecules that can contain hundreds of 'R' groups overall

place. The hormone, 'insulin', for example, is a protein that controls the entry of energy-giving glucose into our cells. Protein molecules

see also...

Antibodies, Enzymes, Proteomics

Proteomics

One of the most celebrated scientific enterprises of recent years has been the Human Genome Project, aimed at determining the chemical structure (base sequence) of all our genes. A rough draft of the human genome is complete, although much remains to be done in checking the draft and analysing all the different versions of genes. The main biochemical significance of genes is that they contain the chemical instructions for manufacturing specific protein molecules. So the big challenge after deciphering the human genome is to tackle what has been called the 'human proteome' – the complete set of proteins needed to build a human, including all the variants associated with both illness and good health. Proteomics, the investigation of the human proteome and those of many other species, will be one of the greatest chemical enterprises of the twenty-first century.

Our genome probably codes for between 100 000 and 1 million protein molecules. That number is very uncertain, because many variants can be produced by 'splicing' together the genetic sequence coding for small parts of proteins in different combinations, and by chemical modification of the resulting proteins. To understand each protein we must determine its amino acid sequence; the complex pattern of folding that occurs to give the protein its final 3-D shape; the chemical modifications that may occur to yield a functional protein; and then determine what biological function the protein has. The possible chemical modifications include the addition of phosphate groups, carbohydrate groups, lipid (fatty) groups, hydrocarbon groups, and the incorporation of metal ions and co-enzymes into the protein's structure.

The main incentive to understand the proteome is the simple fact that the chemical activities of proteins underpin just about every aspect of disease and ageing. Fully understanding our proteins will yield many new ways to prevent and treat disease and to slow or reverse some of the ravages of ageing.

see also...

Carbohydrates, Deoxyribonucleic Acid (DNA), Enzymes, Proteins

Quantum Chemistry

The elementary picture of atoms describes them as mini planetary systems, with tiny electrons whirling around the atomic nucleus in the same way as satellites orbit the Earth. This is a useful model for understanding how protons, neutrons and electrons combine to form atoms. It can also be a useful way to visualise the rearrangements of electrons that accompany chemical reactions. It is, however, a very simplistic view of the atom which bears little relationship to what atoms really are, and what electrons, especially, are really doing. A deeper understanding of the nature of atoms and of chemistry comes from the physical theory known as 'quantum mechanics', which, when applied to chemistry, is called quantum chemistry.

The keystone of quantum chemistry is the fact that things that behave in many ways like little particles also exhibit wave-like behaviour, and things that behave like waves also have particle-like characteristics. This split physical personality is known as 'wave-particle duality'. It is a fundamental feature of both matter and electromagnetic radiation such as light. It has very practical consequences in chemistry. It explains, for example, why electrons must occupy specific regions of space called orbitals. If we treat each electron as a wave, the allowed orbitals correspond to 'standing waves' in the same way as musical notes correspond to standing waves in a plucked guitar string. The standard interpretation of what we really mean, when we treat an electron as a wave, says that the intensity of the electron-wave at any point is related to the *probability* of finding the electron behaving in its particle-like manner at that point. With its emphasis on probabilities, quantum chemistry is a statistical theory, rather than the classical deterministic 'clockwork' view of atoms and electrons following predestined paths. Quantum chemistry focuses much attention on the behaviour of electrons, since electron rearrangement is at the heart of chemistry.

see also...

Atoms, Electron Orbitals, Energy

Radioactivity

The isotopes of some atoms have unstable nuclei which are liable to change spontaneously, accompanied by the release of energy or particles known as nuclear 'radiation'. Any isotope of an element that can release radiation is called a *radioactive isotope*. Whether an isotope is radioactive or not depends on the number of protons and neutrons it contains. Carbon-14 is a radioactive isotope of carbon, with six protons and eight neutrons. A much more abundant and non-radioactive isotope of carbon is carbon-12, with six protons and six neutrons.

The main forms of radiation emitted during the decay of radioactive isotopes are alpha particles (α), beta particles (β) and gamma rays (γ). An alpha particle consists of two protons and two neutrons. Since this is identical to the nucleus of a helium atom, alpha particles are represented in nuclide notation as:

$$^4_2 \text{He}$$

When an alpha particle is ejected from a nucleus, the nucleus changes to that of a different element, since it has lost protons. For example:

$$^{222}_{86} \text{Rn} \rightarrow \, ^{218}_{84} \text{Po} + \, ^4_2 \text{He}$$
Alpha particle

A beta particle is a fast moving electron, ejected when a neutron changes into a proton plus the electron of the beta particle. During this change, the nucleus gains a proton so, again, becomes the nucleus of a different atom. For example:

$$^{14}_6 \text{C} \rightarrow \, ^{14}_7 \text{N} + \, ^0_{-1} \text{e}$$
Beta particle

Gamma rays are highly energetic electromagnetic waves, like light, only with a much shorter wavelength. Gamma rays are often released along with alpha or beta particles. Alpha, beta and gamma radiation are dangerous to living things due to their ability to penetrate into cells and cause damage when they hit sensitive chemicals such as DNA. Gamma rays are the most penetrative, being significantly attenuated only by a substantial layer of dense material such as lead.

see also...
Atoms, Isotopes, Sub-atomic Particles

Rate of Reaction

The rate of a chemical reaction is a measure of how fast it is proceeding. It can be formally defined as the change in concentration or absolute amount of any *product formed* in a given time; or the change in concentration or absolute amount of any *reactant used* in a given time. A simple example will clarify this definition. When calcium carbonate reacts with hydrochoric acid a solution of calcium chloride and bubbles of carbon dioxide gas are produced. Some valid ways to measure the rate of this reaction are:

- mass of carbon dioxide formed per second
- change in concentration of calcium chloride per second
- mass of calcium carbonate consumed per second
- change in concentration of hydrochloric acid per second

Reaction rates are very important to us, because they determine how quickly any chemical reaction we are interested in, or rely on, will proceed. Industrial chemists need to control reaction rates in order to produce their products quickly, but also in a safe manner. Life depends on chemical reactions proceeding at appropriate rates. Medicines must dissolve and take effect at appropriate rates, and so on.

Some key factors that influence the rates of chemical reactions are temperature, concentration, the 'lump size' of solid reactants and the presence of catalysts or inhibitors. Increasing the temperature of a reaction increases its rate of reaction, while cooling slows reaction rates. This is because increasing temperature increases the energy and the frequency of the collisions between particles that initiate reactions. Increasing the concentration of a reactant increases the rate of a reaction, because it makes the collisions between particles more likely, and therefore more frequent. Increasing the surface area of a solid reactant, by grinding a lump down into a powder, for example, increases the rate of reaction because it exposes a greater surface area on the solid reactant.

see also...

Catalyst, Chemical Reaction, Equilibrium

Reduction

If a metal ore such as copper oxide (CuO), is converted into copper metal, this has traditionally been described as a process of reducing the ore down to its pure metal. When chemists became able to examine what happens during such 'reduction' processes, they found that positively charged metal ions gain electrons to form metal atoms. For example:

$$Cu^{2+} + 2e^- \rightarrow Cu$$

This explains the rather confusing fact that to a chemist the word 'reduction' actually means a *gain* of electrons.

Reduction (the gain of electrons) is the complement of oxidation (the loss of electrons). Reduction and oxidation processes occur together in 'redox' reactions. Redox reactions can be set out in an equation scheme showing the oxidation and reduction processes, and overall redox reaction. This is shown, below, for the redox reaction between calcium metal and oxygen gas:

The ion-electron equation for the oxidation of calcium (top line) must be multiplied by two when combining both half-equations, since each oxygen molecule reacts with four electrons. In this reaction the oxygen atoms are reduced while the metal atoms are oxidised to metal ions.

So the following comprehensive definition for reduction is also used: *Reduction is the gain of electrons and/or the addition of hydrogen and/or the removal of oxygen.* This definition explains why the conversion of a carboxyl group (COOH) to an aldehyde group (CHO), is classified as a reduction; as is the conversion of the aldehyde group into an alcohol group (CH_2OH). In the first step an oxygen atom is removed, while in the second hydrogen atoms are added. Conversely, the reverse sequence, changing an alcohol group, into an aldehyde group, then a carboxyl group, involves two oxidation steps.

$Ca \rightarrow Ca^{2+} + 2e^-$	Oxidation
$O_2 + 4e^- \rightarrow 2O^{2-}$	Reduction
$2Ca + O_2 \rightarrow 2CaO$	Redox

see also...

Oxidation

Ribonucleic Acid (RNA)

Ribonucleic acid (RNA) occupies a central role in the chemistry of life. Although it is less famous than the closely related chemical deoxyribonucleic acid (DNA), RNA may actually be the chemical that explains how and why we exist. It may be the chemical that took the first crucial steps in the origin of life.

The key difference between RNA and DNA is that RNA carries an extra oxygen atom on the sugar molecules within the repeating sequence of sugar and phosphate groups that form the 'backbone' of these 'nucleic acid' molecules. Like DNA, RNA molecules have differing sequences of bases strung out along the sugar-phosphate backbone. In living things, three crucial types of RNA are found, known as messenger RNA (mRNA), transfer RNA (tRNA) and ribosomal RNA (rRNA). *Messenger RNA*, a copy of the DNA of a gene, carries the information stored in a gene to the parts of a cell called *ribosomes*, where the genetic information is decoded into the amino acid sequence of a protein molecule. *Ribosomal RNA* molecules form parts of the ribosome. Transfer *RNA* molecules bring the appropriate amino acids to the ribosome for them to be linked into a protein chain. So RNA molecules forge the crucial chemical link between the base sequence of the DNA of our genes and the amino acid sequence of our proteins. In this sense RNA is literally 'central' to life.

In recent years RNA has been found, not only to carry genetic information, but also to be able to act as a biological catalyst – a role previously attributed only to the proteins we call enzymes. Some RNA molecules have also shown signs of being able to catalyse their own replication. This raises the possibility that RNA – a mere chemical – may be capable of the three main features of life: the ability to replicate, to store genetic information and to catalyse chemical reactions. This has led to great interest in the idea that all life may have evolved from a population of spontaneously assembled self-replicating RNA molecules.

see also...

Deoxyribonucleic Acid (DNA),
Enzymes, Proteins

Sanger, Frederick

Some of the most interesting and influential chemical research of the twentieth century was focused on the chemistry of life (biochemistry). At the heart of biochemistry are two classes of giant polymers, the proteins (including all enzymes) and the nucleic acids (including the DNA of our genes). The English biochemist, Frederick Sanger, won two Nobel Prizes for his work in determining the structure of proteins and nucleic acids. He is one of only four people to have won two Nobel Prizes, and has made an outstanding contribution to our understanding of how chemicals allow life forms to live.

In the 1950s Sanger's research group at Cambridge became the first to determine the entire sequence in which a set of amino acids were linked together to form a protein. They worked on the hormone, insulin, but the protein sequencing techniques they developed were widely applied to many other proteins. For this work, Sanger received the Nobel Prize in Chemistry in 1958.

Having devised ways to determine the sequence of amino acids in proteins, Sanger turned his attention to the sequence of nucleotides within nucleic acids such as DNA. Differences in the nucleotide sequence (also the base sequence) of our DNA underpin the differences among different species, different people, and different forms of genes associated with good health or with disease. Sanger's research team devised clever chemical procedures to decipher nucleotide sequences. This allowed them, in the 1970s, to work out the complete nucleotide sequence of the DNA of a very simple virus. This was the first step in developing the techniques that allowed the Human Genome Project to reveal the entire nucleotide sequence of the human genome, a task now completed in draft form, but still being refined and interpreted. Sanger's crucial role in the early work on nucleic acid sequencing was rewarded with the 1980 Nobel Prize in Chemistry.

see also...

Deoxyribonucleic Acid (DNA),
Enzymes, Proteins, Proteomics

Semiconductors

Metals conduct electricity readily because their 'metallic bonding' frees the outer electrons of the metal atoms, so that they can flow through the metal as an electric current when opposite charges are applied to each end of the metal. Non-conductors have no such mobile electrons available. Semiconductors are materials which do not conduct electricity readily, but can be made to do so if just a little energy is applied to make that happen. This allows the semiconductors to be fabricated into the transistors, diodes and so on of the microelectronics industry. The element that lies at the heart of this industry is silicon. Pure silicon is a very poor conductor of electricity. If it is heated, however, the additional heat energy can give some electrons the energy needed to become mobile, and carry a small electric current. This indicates that pure silicon is an 'intrinsic semiconductor', meaning that without modification it can conduct electricity poorly under some conditions.

The properties of silicon become much more useful if just a few of the silicon atoms are replaced with arsenic atoms. This is called 'doping' of the silicon with arsenic. Each arsenic atom carries one more outer electron than a silicon atom, making the doped silicon electron-rich. It therefore conducts electricity more readily than pure silicon, although still not as readily as a metal. An alternative method of doping is to use an element such as *boron*, which has one less outer electron than silicon. This leaves positively charged 'holes' in the structure, which encourage the electrons of silicon to move into them. This creates new 'holes' where the moving electrons have come from, so other electrons jump into these, and the continuation of this process encourages a flow of electric charge, i.e. an electric current. Silicon doped with an electron-rich material is called 'n-type silicon' (where n stands for negative); while silicon doped with electron-deficient material is 'p-type' (p for positive). Microelectronic circuits can be built from 'chips' or 'wafers' of silicon with n-type and p-type parts arranged in complicated ways to control the flow of electrons around the circuit.

see also...

Metals

Smog

The term smog was originally used to describe the mixture of smoke and fog that became a pollution problem in London in the early 1900s. The smoke was a mixture of soot and other particulates released from the burning of coal. It was made more damaging by the presence of sulphur dioxide (SO_2) formed from sulphur in the coal. Many cities around the world now experience another form of smog, called *photochemical smog*, generated by the action of sunlight on the modern mix of urban atmospheric pollution. Automobile exhausts make the greatest contribution to urban atmospheric pollution. These exhausts are composed mainly of carbon dioxide (CO_2) and water (H_2O), which do not contribute to smog; but they also contain small quantities of nitrogen monoxide (NO) and carbon monoxide (CO). The nitrogen monoxide can react with oxygen in the air to form nitrogen dioxide (NO_2). On exposure to bright light, the nitrogen dioxide splits into nitrogen oxide again, plus free oxygen atoms (O), which then react with oxygen molecules to form ozone (O_3). Ozone at ground level is a serious pollutant which irritates our mucous membranes causing respiratory and other problems. The nitrogen dioxide formed in the processes above can react with water to form nitric acid (HNO_3). The nitric acid, ozone and other typical urban pollutants such as sulphur dioxide (SO_2) and a wide variety of organic compounds form a potent chemical mixture. This mixture is a pollution problem even in the absence of sunlight, but intense sunlight initiates a complex series of reactions that converts the mix into highly irritating photochemical smog. Many of the reactions involved occur most quickly on the surface of 'particulates' – tiny lumps of solid such as those formed by the incomplete combustion of diesel fuel. At certain times of the day, and in certain weather conditions, smog is a problem in virtually every busy city.

Photochemical smog is a huge problem in hot and sunny cities such as Los Angeles, where it first came to attention.

see also...
Ozone, Particulates

Soaps and Detergents

The chemicals we use as soaps and detergents must achieve one major chemical task – they must convert substances that are not normally soluble in water into a soluble or suspended form that allows them to be rinsed away. Any dirt and stains that are soluble in water will simply dissolve in the water we use for washing, so there is no real challenge with them. The challenge is set by oily, greasy, 'hydrophobic' (water-hating) chemicals.

Soaps and detergents can solubilize these substances because they have a split chemical 'personality'. Their molecules contain a long non-polar hydrocarbon chain (tail) that is soluble in oils and greases, and a short polar hydrophilic (water-loving) head group that is soluble in water. The hydrophobic tail groups bury themselves in oils and greases, then when agitated they lift and surround tiny droplets of the oily substances. The interaction of the exposed hydrophilic portions of the molecules keeps the oil and grease suspended in the water, allowing it to be rinsed away.

Soaps are made by treating the triglyceride molecules of fats or oils with alkalis such as sodium hydroxide or potassium hydroxide. This hydrolyses the ester linkages in the triglycerides and forms the sodium or potassium salts of the released fatty acids. These are purified to form soap, which is usually made more pleasant to use by the addition of perfumes. One problem associated with soaps is that they form a scum or scale when used in so-called hard water, which contains dissolved calcium ions. The fatty acids in soaps form an insoluble precipitate when they combine with these calcium ions. Detergents avoid this problem, by using synthetic molecules to replace the natural fatty acids of soaps, and these molecules do not form insoluble precipitates with calcium ions. Another way to tackle the problem of water hardness is to remove the calcium ions using a commercial 'water-softening' system.

see also...

Enzymes, Fats and Oils, Polar Molecules

Spectroscopy

When light or other forms of electromagnetic radiation pass through a chemical sample some frequencies of the radiation can be absorbed. This occurs because the energy of each photon of the radiation at these frequencies exactly matches the energy needed to cause some change or 'transition' in the chemical. The technique of passing radiation through chemical samples and detecting which frequencies are absorbed is known as '*absorption* spectroscopy'. It can reveal a great deal about the structure of the chemicals, or can simply identify which chemicals are present in a sample. A similar technique is '*emission* spectroscopy', which involves detection of the light or other electromagnetic radiation emitted when a chemical 'relaxes' from a high energy state into a lower energy state.

In UV–Visible spectroscopy, electromagnetic radiation in the ultraviolet (UV) to visible range of frequencies, is used to induce electrons in atoms and molecules to move between orbitals of different energies. In infra-red (IR)

spectroscopy, infra-red radiation is used to induce molecules to change their modes of vibration, involving the way in which bonds between atoms flex, stretch and compress. These methods of spectroscopy are commonly used by chemists to determine which chemicals are in a sample.

A somewhat more subtle form of spectroscopy is Nuclear Magnetic Resonance spectroscopy (NMR). This relies on the fact that certain nuclei, including those of hydrogen atoms, behave like tiny spinning magnets that will align themselves with an external magnetic field. Electromagnetic radiation in the radio-wave range can be used to flip the nuclei over into the higher energy orientation opposing the external magnetic field. The precise frequencies of radio waves required to do this for atoms in different chemical groups can yield a great deal of information about the structure of the chemicals being studied.

> ### see also...
> *Atoms, Electron Orbitals, Energy*

States of Matter

Nobody fully understands what matter – the 'stuff' of the universe – really is. We have learned how to describe and predict its behaviour, however, in very sophisticated and useful ways. We have found that matter often behaves as if composed of tiny bits, called particles, such as atoms, molecules and ions. We also know, however, that some aspects of matter closely match the behaviour of waves. When we deal with matter in the everyday world, we find that the chemicals concerned generally exist in one of the three basic states – the solid, liquid or gas states. Solids are hard or reasonably firm substances that retain their shape due to the strong forces of attraction among their atoms, molecules or ions. Liquids have weaker forces of attraction among their atoms, molecules or ions, so their shape adjusts to fit the bottom of whatever container they are put into, and they can flow from place to place. In gases, the atoms, molecules or ions are very far apart, moving almost independently with only very weak forces of attraction among them. Gases expand to occupy the entire volume of any container they are placed in. The particles of gases are moving much more quickly than those of liquids, while those in liquids are moving more quickly than those in the corresponding solids. Hence, a gas is converted into a liquid by cooling, a liquid is converted into a solid by further cooling; and transitions from the solid, to the liquid, then the gas state are achieved by heating. These 'changes of state' are not chemical changes, but are known as physical changes, since the identity of the chemicals concerned does not change. As we heat ice, for example, and change it into liquid water, then water vapour, we are simply changing the average speed of the water molecules and, thus, the strength of the intermolecular forces of attraction. Under certain conditions of temperature and pressure chemicals may change directly from the solid to the gas state, and vice versa, without passing through the liquid state. This process is called *sublimation*.

see also...

Atoms, Intermolecular Forces, Quantum Chemistry

Sub-atomic Particles

Atoms are made of three sub-atomic particles known as protons, neutrons and electrons. These combine in varying numbers to form the atoms of all the 91 naturally occurring and the 20 or so human-made elements. The protons and neutrons are bound together within the nucleus of an atom, while the electrons move around the nucleus in electron orbitals.

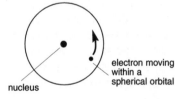

electron moving within a spherical orbital

nucleus

The essential characteristics of the sub-atomic particles are as follows:

	Relative mass	Charge	Location
Proton	1	+1	Nucleus
Neutron	1	0	Nucleus
Electron	Almost zero	−1	Electron orbitals

The electron is believed to be an indivisible fundamental particle. The much more massive protons and neutrons, on the other hand, are composed of smaller particles called 'quarks'. This inner structure allows a neutron to be transformed into a proton and a fast moving electron (known as a β-particle) during the radioactive decay of some atomic nuclei. Another form of radiation, the α-particle, is a cluster of two protons and two neutrons that is ejected from the larger nuclei of some radioactive atoms.

During chemical reactions, electrons can be lost or gained from atoms to form positive or negative ions. The electrons initially belonging to individual atoms can also become shared among several atoms during the formation of the covalent bonds that hold molecules together. In molecules, the electrons occupy orbitals surrounding two or more nuclei, rather than the single nucleus of an atom. The protons and neutrons within the nuclei of atoms are not subject to any changes in chemical reactions.

see also...

Atoms, Covalent Bonds, Electron Orbitals, Ionic Bonds, Molecules, Radioactivity

Substitution Reactions

Any reaction in which a particular atom or group of atoms in a reactant becomes replaced by a different atom or group of atoms is called a substitution reaction. In the reaction below, for example, the chlorine atom (Cl) is replaced by a hydroxyl group (OH).

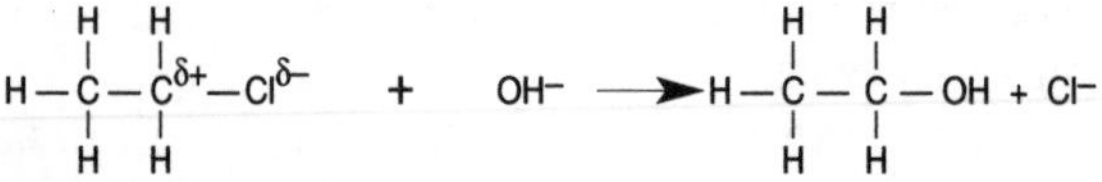

electrophile – a chemical group that is attracted to negative charge.

Substitution reactions are involved in a great many natural and industrial reactions involving organic (carbon-based) chemicals. They are one of the standard processes used by chemists

$$H-\underset{\underset{H}{|}}{\overset{\overset{H}{|}}{C}}-\underset{\underset{H}{|}}{\overset{\overset{H}{|}}{C}}{}^{\delta+}-Cl^{\delta-} \quad + \quad OH^- \quad \longrightarrow H-\underset{\underset{H}{|}}{\overset{\overset{H}{|}}{C}}-\underset{\underset{H}{|}}{\overset{\overset{H}{|}}{C}}-OH + Cl^-$$

The negatively charged OH^- group is a 'nucleophile', meaning a chemical group that is attracted to positive charge. The reaction is initiated by the attraction between the OH^- nucleophile and the slightly positive ($\delta+$) charge on the carbon atom of the

to alter the functional groups contained within molecules. Some substitution reactions are initiated by light. In the process shown below, for example, light energy breaks the Cl–Cl bond to initiate the substitution reaction:

$$H-\underset{\underset{H}{|}}{\overset{\overset{H}{|}}{C}}-H \quad + \quad Cl-Cl \quad \xrightarrow{\text{light}} \quad H-\underset{\underset{H}{|}}{\overset{\overset{H}{|}}{C}}-Cl \quad + \quad H-Cl$$

C–Cl polar covalent bond. Since the nucleophile acts as an attacking group initiating the reaction, this is known as nucleophilic substitution. The Cl^-, which is substituted during the reaction, is called the leaving group. Another form of substitution reaction is electrophilic substitution, which is initiated by attack by an

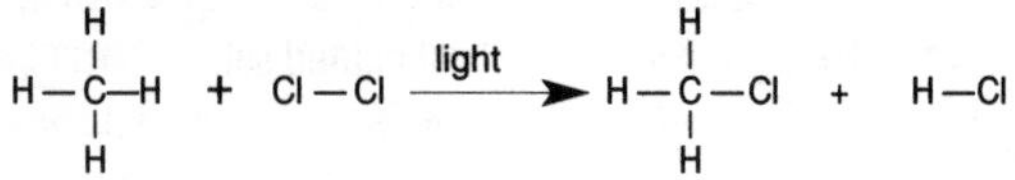

see also...

Electrophiles and Nucleophiles, Functional Groups, Polar Covalency

Superconductors

In 1911 the Dutchman Heike Kamerlingh Onnes discovered that when the metallic element mercury was cooled below 4 K (–269 °C), close to Absolute Zero (0 K, minus 273 °C) it lost all electrical resistance and became a 'superconductor'. Onnes, of Leiden University, was awarded the 1913 Nobel Prize in Physics for this discovery. The phenomenon of superconductivity was later found to be shared by a further 26 of the metallic elements, and by hundreds of alloys (homogeneous mixtures of elements) and compounds.

From the 1980s onwards, scientists have been discovering a wide range of so-called 'high temperature superconductors'. These still have to be made very cold, by everyday standards, to become superconducting, but with 'transition temperatures' in the region of 30–140 K (minus 243 to minus 133 °C), these temperatures are attained considerably more easily than 4 K. Most of the high temperature superconductors are ceramic materials containing copper and oxygen alongside less familiar elements such as lanthanum, yttrium, or barium. The first high temperature superconductor not containing copper was discovered in 1999. One advantage with these materials is that they can be cooled to their superconducting temperatures using liquid nitrogen, rather than the much more expensive liquid helium needed for reaching the lowest temperatures.

In 2001 a carbon-based 'plastic' superconductor was discovered. Although it had a very low transition temperature of 2.6 K (minus 271.4 °C), it may lead to the development of similar superconducting plastics with much higher transition temperatures. This is a big prize to aim for, since these might serve as very cheap and easily fabricated superconducting plastics for use in computing and many other aspects of high technology.

The advantages of materials that conduct electricity without any resistance have already led to superconductors being used in many technological applications.

see also...

Elements, Metals

Further Reading

Books

Atkins, P.W. *Atoms, Electrons and Change* (Scientific American Library, 1990).
Atkins, P.W. *The Periodic Kingdom – a journey into the land of the chemical elements* (Perseus Books Group, 1997).
Conoley, C., Hills, P. *Chemistry* (Collins Educational, 1998).
Emsley, J. *Consumer's Good Chemical Guide* (W.H. Freeman, 1994).
Emsley, J. *Molecules at an Exhibition – portraits of intriguing materials in everyday life* (Oxford University Press, 1999).
Emsley. J. *The Elements* (Oxford University Press, 1998).
Hill, G. *Chemistry Counts* (Hodder & Stoughton, 1995).
Hill, J.W. and Kolb, D.K. *Chemistry for Changing Times* (Prentice Hall, 1998).
Hunt, A. *The Complete A-Z Chemistry Handbook* (Hodder & Stoughton, 1998).
Joesten, M.D. and Wood, J.L. *World of Chemistry* (Saunders, 1991).
Kelter, P.B., Carr, J.D. and Scott, A. *Chemistry – a world of choices* (WCB McGraw-Hill, 2002).
Scott, A. *Molecular Machinery – the principles and powers of chemistry* (Basil Blackwell, 1989).
Selinger, B. *Chemistry in the Marketplace* (Harcourt, Brace & Co., 1998).
Stanitski, C.L., Eubanks, L.P., Middlecamp, C.H. and Stratton, W.J. *Chemistry in Context – applying chemistry to society* (American Chemical Society / McGraw-Hill, 1994).

Web Sites

American Chemical Society: http://www.acs.org/
Chemdex: http://www.chemdex.org/
ChemIndustry.com: http://www.neis.com/
Chemistry Societies Network: http://www.chemsoc.org/
Periodic Table on the WWW: http://www.webelements.com/
The Alchemist online magazine: http://www.chemweb.com/alchemist

Also available in the series

TY 101 Key Ideas: Astronomy	Jim Breithaupt	0 340 78214 5
TY 101 Key Ideas: Buddhism	Mel Thompson	0 340 78028 2
TY 101 Key Ideas: Business Studies	Neil Denby	0 340 80435 1
TY 101 Key Ideas: Ecology	Paul Mitchell	0 340 78209 9
TY 101 Key Ideas: Economics	Keith Brunskill	0 340 80436 X
TY 101 Key Ideas: Evolution	Morton Jenkins	0 340 78210 2
TY 101 Key Ideas: Existentialism	George Myerson	0 340 78152 1
TY 101 Key Ideas: Genetics	Morton Jenkins	0 340 78211 0
TY 101 Key Ideas: Linguistics	Richard Horsey	0 340 78213 7
TY 101 Key Ideas: Philosophy	Paul Oliver	0 340 78029 0
TY 101 Key Ideas: Physics	Jim Breithaupt	0 340 79048 2
TY 101 Key Ideas: Politics	Peter Joyce	0 340 79961 7
TY 101 Key Ideas: Psychology	Dave Robinson	0 340 78155 6
TY 101 Key Ideas: World Religions	Paul Oliver	0 340 79049 0